What Does the Earth Sound Like?

159 ASTOUNDING SCIENCE QUIZZES

EVA EVERYTHING

ECW PRESS

Published by ECW Press
2120 Queen Street East, Suite 200, Toronto, Ontario, Canada M4E 1E2
416.694.3348 / info@ecwpress.com

LIBRARY AND ARCHIVES CANADA CATALOGUING IN PUBLICATION

Everything, Eva
What does the earth sound like? : 159 astounding
science quizzes / Eva Everything.

ISBN 978-1-77041-009-1
ALSO ISSUED AS:
978-1-55490-910-0 (PDF); 978-1-55490-979-7 (EPUB)

1. Science—Miscellanea. I. Title.

Q173.E93 2011 500 C2010-906724-X

Editor: Crissy Boylan
Cover and text design: Tania Craan
Cover images: Earth © Andrey Prokhorov;
violin © Juanmonino; drumming © Allen Johnson;
Om © tjasam; spinning top © LiciaR (all iStockPhoto.com)
Author photo: Biserka
Typesetting: Mary Bowness
Production: Troy Cunningham
Printing: Friesens 1 2 3 4 5

This book is set in Helvetica and Fairfield

The publication of *What Does the Earth Sound Like?* has been generously supported by the Government of Ontario through Ontario Book Publishing Tax Credit, by the OMDC Book Fund, an initiative of the Ontario Media Development Corporation, and by the Government of Canada through the Canada Book Fund.

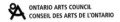

Canada
ONTARIO ARTS COUNCIL
CONSEIL DES ARTS DE L'ONTARIO

PRINTED AND BOUND IN CANADA

ECW PRESS
ecwpress.com

FSC
www.fsc.org
MIX
Paper from responsible sources
FSC® C016245

FOR MY FATHER,
AND FOR YOU

Acknowledgements

People sometimes ask me: how do you come up with these questions? My inspiration comes from the work of brilliant scientists who tackle the questions no one thought to ask before and who freely reveal the answers. A shout-out goes to all of them, both dead and living, with special thanks to Dr. Chuck Gerba (a.k.a. The Germinator), who is alive and well and whose studies will blow your mind and possibly even horrify you; Dr. Brian Wansink whose experiments with food never cease to amaze and amuse; and Dr. Curtis Ebbesmeyer who studies ocean currents by tracking things lost at sea, including plastic yellow duckies. I'm also grateful to my family and friends for their love and support, the Ontario Arts Council Writers' Reserve, and the wonderful team at ECW Press who continue to pursue the impossible dream by publishing books.

Table of Contents

Introduction

This quiz book is actually a "book" disguised as a quiz book. Let me explain. I could have written a prose book on the same subjects covered here, but would it be as much fun to read as a quiz book? Would it challenge your imagination and make you wonder, even about things you probably take for granted? Most quiz books are bare-bones questions and answers. They're fun in a superficial kind of way, but it's impossible to learn anything meaningful without its context. A book written in prose gives you context but doesn't challenge your brain in the same way quizzes do. This book is a hybrid of sorts, combining the best aspects of both formats — so you get both context and fun on the same page.

The world (and I mean that in the broadest sense of the word) is endlessly fascinating. Some people believe that all the greatest discoveries have already been made. Oh, really? The universe is bigger than we can imagine, and so is the potential for great ideas. New discoveries are made each and every day. Even the things we think we know are constantly being revised. So much so that, by the time, you read this, some of the content could be yesterday's news. Don't blame me — blame the blistering pace of scientific discovery and human innovation!

If you're used to a traditional quiz book format, you're in for a pleasant change. No more thumbing to the back of the book for the answers! To see the answer, all you have to do is turn the page and flip the book. Ready to move on to the next question? Just flip the book back.

I hope you enjoy reading this book as much as I enjoyed writing it. If it inspires you to delve more deeply into the subjects covered in these pages, by all means, follow your curiosity. You never know where it will lead. In a letter to his biographer, Albert Einstein claimed, "I have no special talent. I am only passionately curious." I'm no Einstein, but I'm curious too. What about you?

Eva Everything
www.thebraincafe.ca

Q

True or false? Humans have the biggest brain compared to their body size. Did you say true? That's correct! Did you say false? Also correct! How can that be? Simple. At one stage of life your brain is absolutely the biggest compared to body size, but at other stages it's not.

At what stage of life is your brain the biggest compared to your body?

a) baby/toddler
b) teenager
c) young adult
d) mature adult

The Biggest Brains

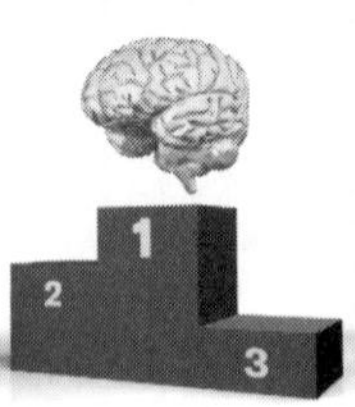

A

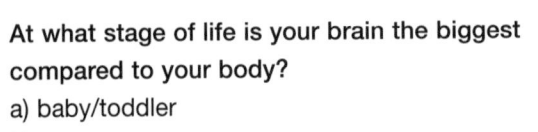

At what stage of life is your brain the biggest compared to your body?

a) baby/toddler
b) teenager
c) young adult
d) mature adult

CORRECT ANSWER:
a) baby/toddler

At about 10% of your body weight, your brain was huge. Adult brains are about 2% of body weight, but mouse, squirrel monkey, and hummingbird brains are more than 3%. Smaller animals tend to have bigger brains relative to body size, but thanks to our big baby brains, we're still number one.

Growing a Brain

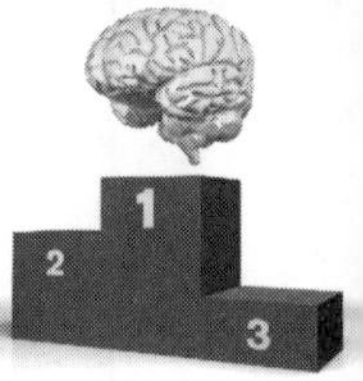

You started growing a brain soon after conception. In the first couple of months, your brain had a big growth spurt and raced to make all kinds of brain cells, including neurons.

How many neurons did your brain grow in a minute?

a) 250

b) 2,500

c) 25,000

d) 250,000

Growing a Brain

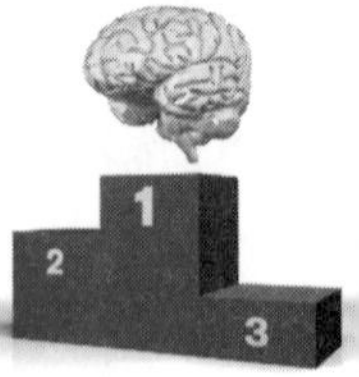

How many neurons did your brain grow in a minute?

a) 250
b) 2,500
c) 25,000
d) 250,000

CORRECT ANSWER:
d) 250,000

That works out to more than 4,000 neurons per second! By the time you were three years old, your brain was 90% of its adult size. After that, your body started growing into that big brain of yours.

Biological Supercomputer

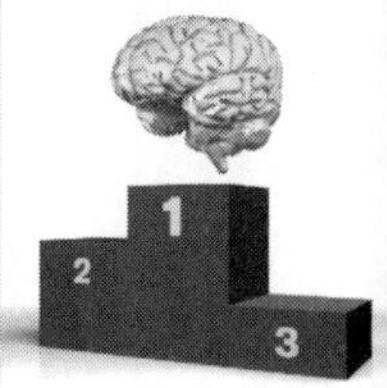

Q

The best supercomputer on the planet is in between your ears. Even when you're asleep, it's online running all your systems and processing vital information. Brain cells called neurons are the connections within your biological computer.

How many neurons are there in your brain?
As many as all the . . .

a) stars in the Milky Way
b) telephone numbers in the world
c) websites on the internet
d) world's human population

Biological Supercomputer

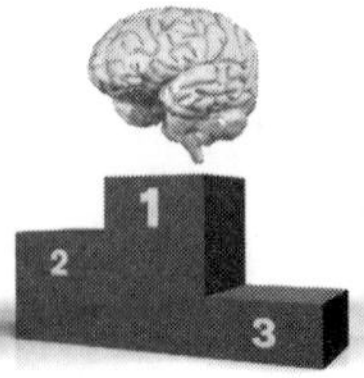

A

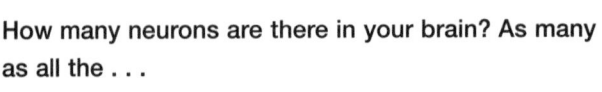

How many neurons are there in your brain? As many as all the . . .

a) stars in the Milky Way
b) telephone numbers in the world
c) websites on the internet
d) world's human population

CORRECT ANSWER:

a) stars in the Milky Way

There are about a hundred billion neurons packed into your compact biological supercomputer. They can make or break a million connections per second. Man-made supercomputers are faster, but beyond that there's no comparison. After all, can supercomputers come up with ideas, appreciate chocolate, or enjoy a good book?

Q

Are you and your pet alike? In an online survey, 40% of respondents who'd had their pet for seven years or more said its personality was like their own. British psychologist Richard Wiseman (Quirkology.com) asked people to rate themselves and their pets on traits such as happiness, intelligence, independence, and sense of humour. According to the survey . . .

Which pets have the best sense of humour?

a) birds

b) cats

c) dogs

d) fish

The Most Hilarious Pet

Which pets have the best sense of humour?

a) birds

b) cats

c) dogs

d) fish

CORRECT ANSWER:

c) dogs

More than 60% of the dog owners thought that their pooch had a good sense of humour. Surprisingly, almost as many fish owners claimed to have hilarious pets. Perhaps the fish owners, and not the fish, were the ones with a sense of humour. Cats, horses, and birds rounded out the top five. Reptiles were at the bottom of the list. Apparently, they have no sense of humour at all.

The Happiest Pet People

Forty percent of the dog owners in Dr. Richard Wiseman's online survey described themselves as fun loving, compared to only 2% of the reptile owners. However, reptile lovers did rate themselves the most independent. Cat owners rated themselves the most dependable and emotionally sensitive.

Which pets had the happiest owners?

a) birds

b) cats

c) dogs

d) fish

Q

The Happiest Pet People

Which pets had the happiest owners?

a) birds

b) cats

c) dogs

d) fish

CORRECT ANSWER:

d) fish

About 37% of the people with a fish strongly agreed to being happy. Cat people were the next happiest at 24%, followed closely by dog people at 22%. Would you be happier if you had a fish? Maybe it's the other way around. If you were happier, you might want to keep a fish.

Blub-bye

The sad truth is that fish often don't live as long as many other pets. No one knows how much pain they feel, but seeing a finny friend slowly going belly up can be distressing for a human. If it's the right thing to do . . .

Q

What's the most humane way to euthanize a pet fish?

a) buy it a drink

b) feed it to a bigger fish

c) flush it down the toilet

d) put it in the freezer

Blub-bye

What's the most humane way to euthanize a pet fish?

a) buy it a drink

b) feed it to a bigger fish

c) flush it down the toilet

d) put it in the freezer

CORRECT ANSWER:

a) buy it a drink

If it's time for the final farewell, sliding your finny friend into a container holding one part alcohol and four parts water is a humane solution. The fish gets a painless, numbing buzz and passes out forever. Blub-bye, li'l buddy. RIP.

How much G-force does it take to black out? Even after World War II, no one was sure. In order to design a stronger, more lightweight crash helmet for fighter pilots, Royal Canadian Air Force scientists needed to know. So, they came up with a unique top secret experiment to find out how many Gs it took to lose consciousness.

What did they do?

a) banged their heads against steel plates
b) crashed cars into brick walls
c) flew high enough, fast enough, to pass out
d) jumped out of planes without parachutes

What did they do?

a) banged their heads against steel plates
b) crashed cars into brick walls
c) flew high enough, fast enough, to pass out
d) jumped out of planes without parachutes

CORRECT ANSWER:

a) banged their heads against steel plates

The scientists attached devices that measure acceleration to their own foreheads and banged them into a steel plate as hard as they could. The brain-bashing reached forces of up to 10 Gs, more than enough to black out. (Most people black out at between 4 and 5 Gs.) When their commanding officer saw what they were up to, the top secret head-banging experiment came to an abrupt end.

The First G-Suit

Because G-suits hadn't been invented yet, pilots were blacking out during high-speed maneuvers. In the 1930s, Dr. Wilbur R. Franks had an idea that he hoped would keep pilots from passing out while pulling extreme Gs. But before he made a suit for humans, the Canadian scientist tested the concept on animals.

What did he do to test his idea for a G-suit?

a) dressed chimps in diving suits
b) molded foam rubber suits to fit beavers
c) put mice in water-filled condoms
d) tethered dogs inside a plastic bubble

The First G-Suit

What did he do to test his idea for a G-suit?

a) dressed chimps in diving suits

b) molded foam rubber suits to fit beavers

c) put mice in water-filled condoms

d) tethered dogs inside a plastic bubble

CORRECT ANSWER:

c) put mice in water-filled condoms

The mice tolerated up to an incredible 240 Gs inside their water-filled condoms. Dr. Franks had a fitted, water-filled top secret G-suit made for himself and got into a small training plane. He must have freaked when the pilot pulled 7 Gs and temporarily blacked out because Dr. Franks, who had never flown before, was fully conscious in his water-filled G-suit. His ingenious invention was adapted for the G-suits worn by pilots, astronauts, and cosmonauts today.

Project Habbakuk was a top secret World War II project hidden away in the Canadian Rocky Mountains. The goal was to build an unsinkable aircraft carrier for the British Navy. It had to be more durable than regular aircraft carriers and cheaper to build. If it could be made out of something that wouldn't attract magnetic mines, all the better.

What was used to make the aircraft carrier?

a) concrete

b) cork and latex

c) recycled plastic water bottles

d) water and wood pulp

Project Habbakuk

What was used to make the aircraft carrier?

a) concrete

b) cork and latex

c) recycled plastic water bottles

d) water and wood pulp

CORRECT ANSWER:

d) water and wood pulp

Fifteen men spent months building a model out of a slurry of water and wood pulp called Pykrete. The frozen mixture is stronger than plain ice, takes longer to melt, and can be machined like wood or cast like steel. It worked but wasn't practical. Project Habbakuk was cancelled, and the top secret ice ship melted into the waters of Lake Patricia. Its refrigeration system sank to the bottom, and it's no secret that it's still down there.

Q

How about a glass of nice warm pigeon milk? It's rich in fat, protein, and vitamins and is more nutritious than milk from cows or humans. A teaspoon of the stuff contains as much vitamin A as a drop of cod liver oil! It's so nourishing that baby pigeons fed the special milk double their weight in less than 36 hours after hatching.

Where does pigeon milk come from?

a) part of the pigeon's digestive system

b) pores in the skin

c) teats under the wings

d) their behinds

Pigeon Milk

Where does pigeon milk come from?

a) part of the pigeon's digestive system
b) pores in the skin
c) teats under the wings
d) their behinds

CORRECT ANSWER:

a) part of the pigeon's digestive system

Both female and male pigeons make milk from the lining of their crop. Emperor penguins and greater flamingos produce their version of milk too, but in other parts of their digestive tracts. Despite its nutritional value, you might not want pigeon milk in your tea, coffee, or cereal. Its consistency is a bit like earwax.

Milk to Go

Q

When you remove all the water from milk, you're left with milk powder. It's not big with consumers, but food manufacturers put it in everything. From chocolate and cookies to macaroni and cheese mixes, if "milk" is listed in the ingredients, it's probably milk powder.

Who used milk powder first?

a) Allied troops during World War II

b) astronauts

c) Huns

d) Vikings

Milk to Go

Who used milk powder first?

a) Allied troops during World War II

b) astronauts

c) Huns

d) Vikings

CORRECT ANSWER:

c) Huns

More than 1,500 years ago, Attila the Hun's hordes conquered the world on horseback, nourished by mare's milk in the form of milk powder. The Mongols, who were also horseback conquerors, made mare's milk powder too. It's manufactured in some parts of the world to this day. Mare's milk powder is not just for conquerors anymore.

They All Screamed for Ice Cream

Americans are among the world's top ice cream eaters. During World War II, American airmen stationed in Britain missed their favourite frozen treat. They had powdered milk and sugar but no ice cream makers, so they mixed up their ingredients and improvised.

How did they make ice cream?

a) in canisters dropped from high altitudes

b) in natural springs of icy cold water

c) in the tail gunner's compartment of bomber aircraft

d) with dry ice from Oxford University's physics lab

They All Screamed for Ice Cream

How did they make ice cream?

a) in canisters dropped from high altitudes
b) in natural springs of icy cold water
c) in the tail gunner's compartment of bomber aircraft
d) with dry ice from Oxford University's physics lab

CORRECT ANSWER:

c) in the tail gunner's compartment of bomber aircraft

During high-altitude bombing missions, B-17 Flying Fortress bombers sometimes doubled as ice cream makers. The airmen attached canisters of ice cream mix to the tail gunner's compartment. The airplane's movement shook it up nicely as it froze in temperatures that could plunge as low as –60°C (–76°F).

TOUR OF THE SOLAR SYSTEM: THE SPIN

The Spin on the Solar System

Q

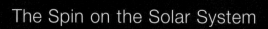

There you are hovering above the solar system, looking down at the sun's north pole. It's a spectacular sight. The sun is huge, making up more than 99% of the mass of the solar system. Even the biggest planets look puny in comparison. Since you can't afford to spend years watching the planets orbit in real time, you fast-forward and scan their motion.

How do the planets orbit the sun?

a) clockwise

b) counterclockwise

c) some clockwise, some counterclockwise

d) They don't. The sun orbits Earth.

The Spin on the Solar System

How do the planets orbit the sun?

a) clockwise

b) counterclockwise

c) some clockwise, some counterclockwise

d) They don't. The sun orbits Earth.

CORRECT ANSWER:

b) counterclockwise

The entire solar system started out as a huge cloud of hot gas that started rotating counterclockwise. It eventually collapsed into a super hot, spinning disc. The sun sucked up most of the disc and the planets formed from the leftovers. They still orbit the sun in the same nearly flat disc, which is called the ecliptic plane.

The Spin on Earth

When Earth was a hot newbie planet 4.5 billion years ago, it was spinning really fast. No one knows exactly how long a day was, but a computer simulation calculated it to be about six hours long. As the planet's spin slowed, the days got longer. Some scientists think that by 620 million years ago, a day was 21.9 hours long.

How do they know that?

a) ancient timepiece built by aliens

b) annual layers of sandstone

c) annual tree rings

d) layers in ice cores

The Spin on Earth

How do they know that?

a) ancient timepiece built by aliens
b) annual layers of sandstone
c) annual tree rings
d) layers in ice cores

CORRECT ANSWER:
b) annual layers of sandstone

The answer may be writ in stone. Based on studies of ancient sandstone layers, some scientists think that 620 million years ago a day was about two hours shorter than it is today. In the past 3,000 years, the Earth's spin has slowed by about 0.05 seconds. To give you an idea of how long that isn't, it takes three or four times longer to blink.

The Fastest Spin

The time it takes for a planet to spin on its axis determines how long its day is. You know how long a day is on Earth, but what do you know about other planets? Their days range from extremely short to incredibly long.

Which planet has the shortest day?

a) Jupiter

b) Mars

c) Mercury

d) Venus

The Fastest Spin

Which planet has the shortest day?

a) Jupiter
b) Mars
c) Mercury
d) Venus

CORRECT ANSWER:
a) Jupiter

The biggest planet in the solar system spins fastest on its axis and has a day that's only 9.9 Earth hours long. A day on Jupiter is short, but a year is 10,499 Jovian days long. If you had a job on the smallest planet, you'd have the longest workday in the solar system. One solar day on Mercury (the time it takes to spin on its axis once) is about two Mercurian years long. You'd get annual coffee breaks.

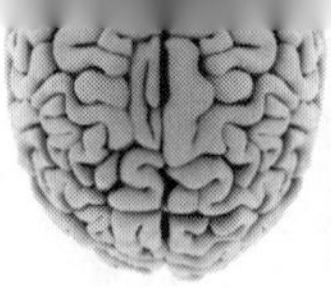

Bigger and Better Brains

You don't have to be a scientist to know that men and women aren't the same. Even their brains are different. Men are generally bigger than women and so are their skulls and brains. Compared to a normal female brain . . .

How much bigger is the male brain on average?

a) 1%
b) 5%
c) 10%
d) 15%

Bigger and Better Brains

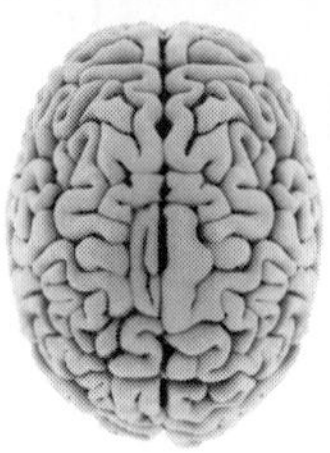

How much bigger is the male brain on average?

a) 1%

b) 5%

c) 10%

d) 15%

CORRECT ANSWER:

c) 10%

Let the gloating begin! Not so fast, guys. Females have smaller brains but more grey matter, which is the thinking part of the brain. That makes them smarter, right? Well, no. Males and females score equally well on intelligence tests. It's sort of like the difference between an equally powerful desktop and a laptop. They don't look the same, but when it comes to what they can do, they're not all that different.

Is Your Brain Shrinking?

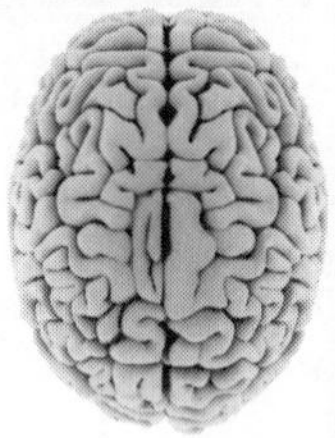

Q

Your brain grows at an amazing pace from the time you're a fetus to your early childhood. By the age of three, your brain's diameter is 95% of its adult size. Once your brain stops expanding, it naturally starts shrinking.

At what age does your brain start to shrink?

a) 20–25

b) 32–36

c) 40–43

d) 60–62

Is Your Brain Shrinking?

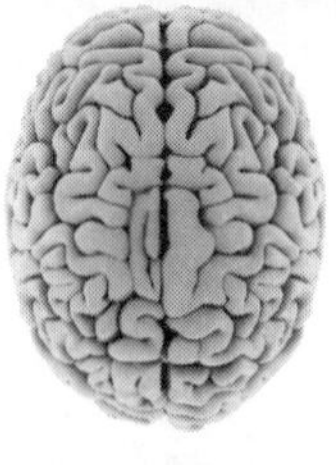

At what age does your brain start to shrink?

a) 20–25
b) 32–36
c) 40–43
d) 60–62

CORRECT ANSWER:
a) 20–25

How much and how fast your brain shrinks depends on your genes and on your sex. There's lots of individual variation but, on average, male brains shrink about 10% per decade. Female brains don't tend to shrivel up nearly as much. One brain expert estimated that, on average, adults lose a neuron per second. It's a good thing we started out with so many.

Groovy Wrinkles

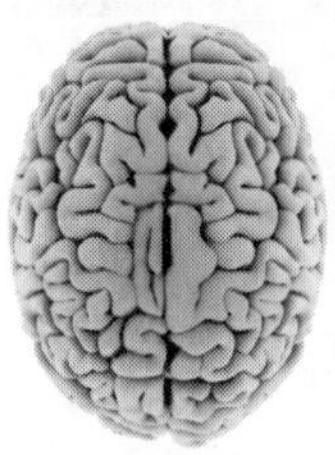

Q

When it comes to your brain, think wrinkles. They ramp up your brainpower by increasing the surface area of your cerebral cortex, or grey matter. If you take a sheet of paper and crumple it up into a ball, it still has the same surface area as when it was flat. Your brain is somewhat like that, with lots of grey matter folded into a small wrinkly package. How big would it be if you smoothed it out? If you flattened out an average-sized cerebral cortex . . .

About how many sheets of letter-sized paper (8.5 x 11 in) would it cover?

a) 1
b) 2
c) 3
d) 4

Groovy Wrinkles

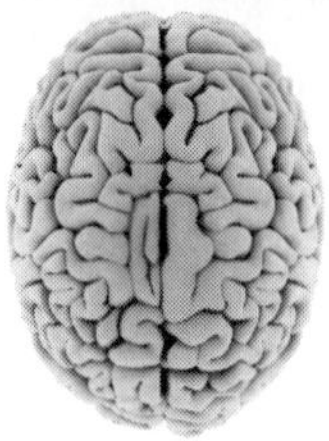

A

About how many sheets of letter-sized paper (8.5 x 11 in) would it cover?

a) 1
b) 2
c) 3
d) 4

CORRECT ANSWER:

d) 4

Flattened and smoothed out, your grey matter would cover roughly four sheets of letter-sized paper. Just like a crumpled sheet of paper, more than two thirds of your grey matter is buried in the grooves of your brain. Your grey matter multi-tasks 24/7, processing attention, perception, thought, language, memory, consciousness, and much more. Pretty amazing, but the best part is that it does all that without you ever having to think about it!

Your Kid Eats Dirt

Q

There's no doubt that kids eat dirt. Small children touch whatever they can reach. Then, every three minutes or so, they probe their noses and mouths. So how much dirt actually goes down the hatch? The Germinator, a.k.a. microbiologist Chuck Gerba, studied the poop of pre-school aged children to calculate how much dirt they ate. He compared it to the dirt you'd find on typical 20 centimetre by 20 centimetre (8 in by 8 in) floor tiles.

How many floor tiles' worth of dirt did the little kids eat?

a) 1–2

b) 2–4

c) 4–6

d) 6–8

Your Kid Eats Dirt

How many floor tiles' worth of dirt did the little kids eat?

a) 1–2

b) 2–4

c) 4–6

d) 6–8

CORRECT ANSWER:

d) 6–8

Dr. Gerba came up with that number by measuring the amount of metal in the children's stool. He already knew how much metal was in the dust the children ingested, and by measuring how much was in their poop, he could calculate how much dirt they'd eaten. When he compared it to dirt on a typical floor tile, The Germinator discovered that the average small child eats six to eight tiles' worth of dirt a day.

Dust Me!

Q

If you're unlucky enough to be the one who does the dusting at home, you know what a frustrating chore it is. No sooner are you finished, than the dust is settling again. It's probably because each and every hour nearly 1,000 dust particles per square centimetre (0.15 sq in) settle in the average home. Where is all the dust coming from?

What's the most abundant thing in household dust?

a) bits of dust mites and their poop
b) fibres from fabrics
c) fungal spores and pollen
d) human and animal skin flakes

Dust Me!

What's the most abundant thing in household dust?

a) bits of dust mites and their poop

b) fibres from fabrics

c) fungal spores and pollen

d) human and animal skin flakes

CORRECT ANSWER:

d) human and animal skin flakes

You'll find all of the above in household dust, but skin flakes can make up 70 to 90% of it. If you don't spend a lot of time at home, your dust might be less flaky. But if you're home a lot and have pets, skin flakes can make up most of your dust. You, your pets, your home, and anything in it all contribute tiny particles to that dust you just can't seem to get rid of.

It Came from Space

Q

There could be a piece of a comet's tail right there in the room with you. Do you see it? Probably not. If it's there, it's the size of a speck of dust. When you open your doors and windows, comet dust or micrometeorites (dust-sized particles of asteroids) from space can land in your home.

About how many tons of space dust fall to Earth per day?

a) 1–2
b) 10–20
c) 100–200
d) 1,000–2,000

It Came from Space

About how many tons of space dust fall to Earth per day?

a) 1–2

b) 10–20

c) 100–200

d) 1,000–2,000

CORRECT ANSWER:

c) 100–200

It's impossible to measure it all, but 100–200 tons per day is a recent estimate. It can be more, or less, depending on what's going on in the solar system. If Earth passes through meteor showers or the dusty wake of a comet, more space dust falls than usual. As well as micrometeorites, space dust contains tiny specks of space junk.

Q

The kiwi is not only the name of a fruit; it's also a flightless, chicken-sized bird native to New Zealand. Kiwis hold the record for producing the largest egg relative to body size. One egg can weigh as much as 20% of the female's body weight. The ovum, or egg, is one of the biggest human cells. It's obviously nowhere near the size of a kiwi egg, but how big is it really?

How big is the human ovum? About the size of a . . .

a) bee hummingbird egg
b) black peppercorn
c) this period.
d) yellow mustard seed

X-Rated Egg

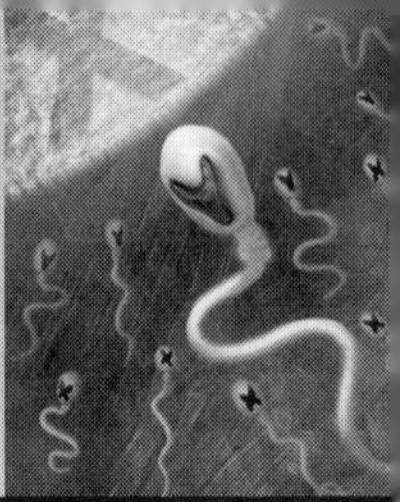

How big is the human ovum? About the size of a . . .

a) bee hummingbird egg
b) black peppercorn
c) this period.
d) yellow mustard seed

CORRECT ANSWER:

c) this period.

A mature human ovum is about one-seventh of a millimetre (0.005 in) across, roughly the size of the period at the end of this sentence. Most mammalian eggs are about the same size, big enough to see with the naked eye. The human ovum always carries an X chromosome from the female.

A

X, Meet Y

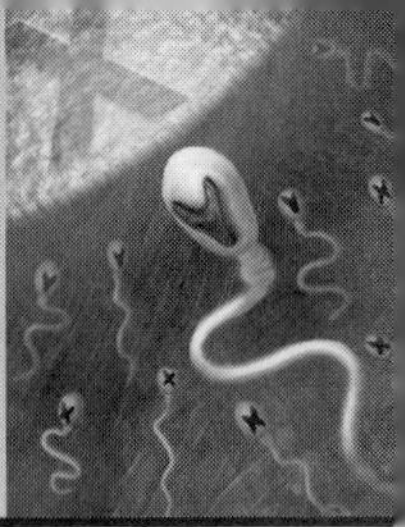

Q

You probably know that sperm are some of the smallest cells in the human body and that ova are some of the biggest. Have you ever wondered how they compare in size? Maybe you've seen animations of frenzied sperm racing toward the egg and jackhammering its surface. Was that scale accurate?

How many sperm could bang their heads into an ovum at the same time?

a) 4
b) 40
c) 400
d) 4,000

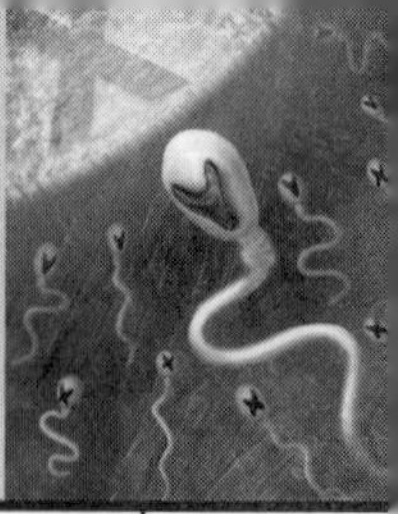

How many sperm could bang their heads into an ovum at the same time?

a) 4

b) 40

c) 400

d) 4,000

CORRECT ANSWER:

d) 4,000

You need a microscope to see a sperm. Its head is roughly the size of a red blood cell. Because sperm have such tiny flat heads and the ovum's surface area is so relatively huge, about 4,000 sperm could theoretically try to penetrate an ovum at the same time. But that doesn't happen in real life because few sperm survive the race to the ovum.

X + Y = Guy

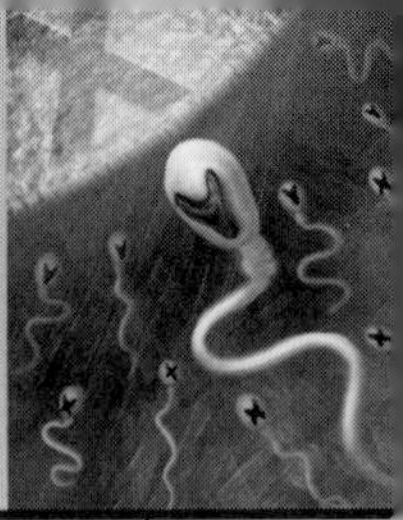

The sperm that fertilizes the ovum determines the sex of the baby. Some sperm carry an X chromosome, for making a girl, and others carry a Y chromosome for making a boy. If the fetus is a boy, a couple of months after conception a chain of chemical reactions kickstarts male development.

How many genes does it take to become male?

a) 1
b) 10
c) 100
d) 1,000

X + Y = Guy

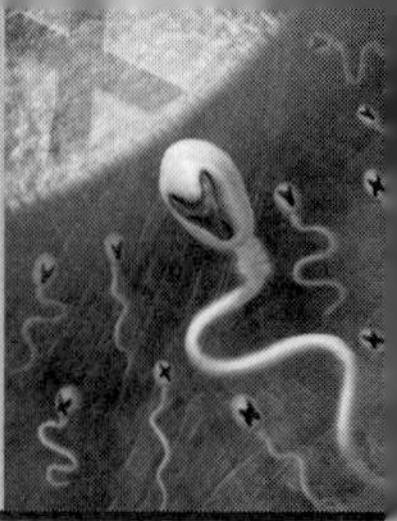

How many genes does it take to become male?

a) 1
b) 10
c) 100
d) 1,000

CORRECT ANSWER:

a) 1

It only takes one gene, called Sry, to start the chemical cascade that leads to maleness. The testicles develop first and start making the male hormone testosterone. It masculinizes the developing body and changes the brain, including its size, structure, and wiring.

Q

Do you know the staggering odds you beat to become you? Of your mother's 400,000 potential eggs, you were the one in a thousand that developed into a fertilizable egg, or ovum. Your sperm beat out as many as 200 million competitors in a life-or-death race. When your ovum and sperm hooked up, various genetic combinations were possible, but only one would produce you.

What were the odds that your combination would come up?

a) 1 in 64 thousand

b) 1 in 64 million

c) 1 in 64 billion

d) 1 in 64 trillion

It Wasn't Easy Becoming You

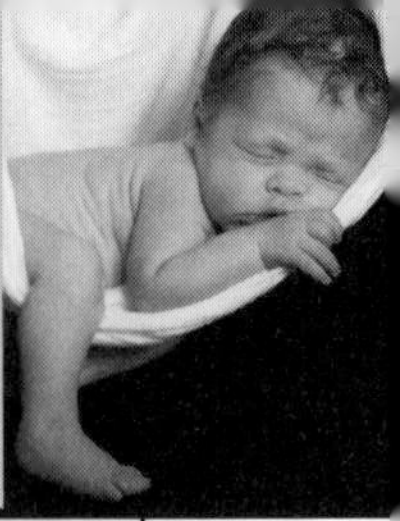

What were the odds that your combination would come up?

a) 1 in 64 thousand
b) 1 in 64 million
c) 1 in 64 billion
d) 1 in 64 trillion

CORRECT ANSWER:
d) 1 in 64 trillion

There's no lottery on Earth with such improbable odds for winning. Of the 64 trillion possible combinations, your special combo won the gene pool. It's highly unlikely that anyone else has ever had, or ever will have, your unique combination — unless you're an identical twin.

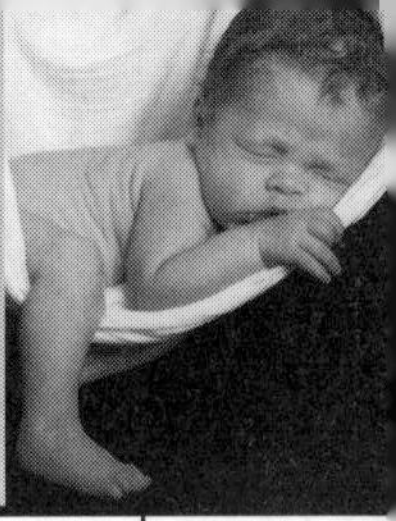

The First-born Advantage?

Why are some people more intelligent than others? Philosophers and scientists have been trying to figure that out for thousands of years. There are lots of ideas and explanations. Maybe you've even heard that being the first-born in your family gives you a boost in brainpower. Well, it's true. Compared to the second-born . . .

How much higher is the IQ of the first-born?

a) 1–2 points

b) 5–10 points

c) 10–15 points

d) 15–20 points

The First-born Advantage?

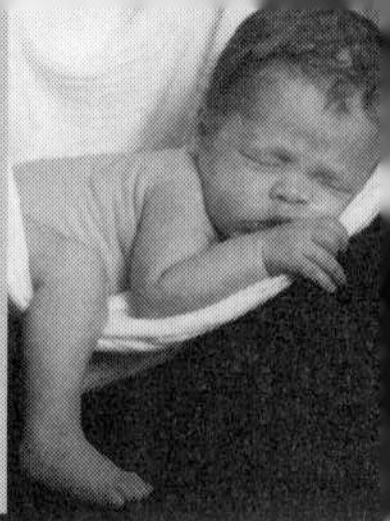

How much higher is the IQ of the first-born?

a) 1–2 points

b) 5–10 points

c) 10–15 points

d) 15–20 points

CORRECT ANSWER:

a) 1–2 points

There usually aren't big differences in the IQ scores of children from the same family. First-borns tend to have a few more IQ points, unless the second-born arrives seven or more years later. Then, it's often reversed. Some experts think that later-born children are more likely to be revolutionary leaders, scientists, or creative types. But first-borns are more likely to live to 100.

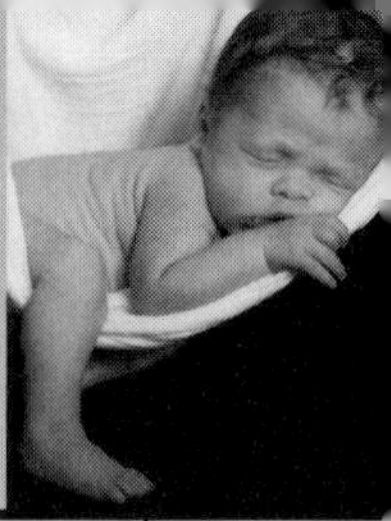

Living to 100

In ancient Rome, most people were dead before they hit 30. A hundred years ago, 50 was pushing it. More and more people are living twice that long these days. Why do some people live to 100? And why are so many of them first-borns? American scientists studied centenarians and discovered something that might double your chance of living a century.

What might double your chance of living to 100?

a) a grandparent who lived to 100

b) being an only child

c) your father was more than 40 years old when you were born

d) your mother was less than 25 years old when you were born

Living to 100

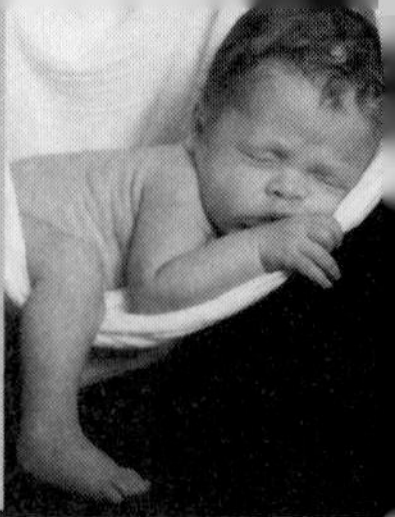

A

What might double your chance of living to 100?

a) a grandparent who lived to 100
b) being an only child
c) your father was more than 40 years old when you were born
d) your mother was less than 25 years old when you were born

CORRECT ANSWER:

d) your mother was less than 25 years old when you were born

The secret of longevity was being born to a young mother, at least in this study. Why? No one's sure, but it could be that young women have better quality eggs and fewer chronic diseases, which makes for healthy babies. So, if you were born to an older mom, does it mean you won't live to 100? The researchers think you still could, thanks to improved hygiene, medicine, and nutrition. They also suggest being as happy and stress-free as possible to enhance your chances.

Q

The experiments and tests that paved the launch pad to space were often risky and sometimes death-defying. Aerospace pioneers put their lives on the line so that future pilots and astronauts would be safe. In 1960, American test pilot Joe Kittinger took *Excelsior III* all the way to the stratosphere, the upper layer of the atmosphere. He was three times higher than the typical altitude of a modern commercial jet.

What was *Excelsior III*?

a) balloon

b) jet pack

c) manned rocket

d) top secret jet fighter

Fly Me to the Stratosphere

What was *Excelsior III*?

a) balloon

b) jet pack

c) manned rocket

d) top secret jet fighter

CORRECT ANSWER:

a) balloon

Excelsior III was a humongous balloon about as tall as a 20-storey building. It was filled with some 85,000 cubic metres (3 million cu ft) of helium. The big balloon took 90 minutes to lift Joe Kittinger, in a gondola rigged with scientific gear, to the edge of space. It was a long and frigid ride all the way up to the stratosphere.

Chilling on the Edge of Space

Wearing a pilot's G-suit and layers of clothing, Joe Kittinger felt the chill as he rose to the edge of space. *Excelsior*'s heating system consisted of plastic water bottles that gave off heat as they froze. Once they'd frozen, Kittinger had to endure temperatures as low as –70°C (–94°F). After he'd reached a record high altitude of 31,333 metres (102,800 ft), he did something that no one else had ever done before.

What did Joe Kittinger do in the stratosphere? He . . .

a) made the first TV broadcast from space

b) smoked a cigarette

c) stepped out of the gondola

d) took telephoto pictures of the moon

Q

Chilling on the Edge of Space

What did Joe Kittinger do in the stratosphere? He . . .

a) made the first TV broadcast from space

b) smoked a cigarette

c) stepped out of the gondola

d) took telephoto pictures of the moon

CORRECT ANSWER:

c) stepped out of the gondola

Framed by the blackness of space, Kittinger plummeted toward Earth. Dropping from high altitudes, bodies fall into a spin, rotating more than 100 times a minute. It's called a lethal flat spin because it can kill you. Kittinger's life depended on the parachute system he was testing. If his parachute didn't open and stabilize his fall, his life was seriously in danger.

Flying Fast and Free

Sixteen suspenseful seconds into Joe Kittinger's death-defying plunge, the stabilizing parachute opened and prevented a lethal flat spin. He set a record for the highest parachute jump and the longest free fall. Some argue that Kittinger wasn't *really* in free fall because he was using the small stabilizing chute. Whatever. Did I mention that he was falling incredibly fast?

How fast was Joe Kittinger falling?

a) almost at the speed of sound

b) commercial jet's typical cruising speed

c) faster than the current land speed record

d) wind speed of a tornado

Q

Flying Fast and Free

How fast was Joe Kittinger falling?
a) almost at the speed of sound
b) commercial jet's typical cruising speed
c) faster than the current land speed record
d) wind speed of a tornado

CORRECT ANSWER:
a) almost at the speed of sound

Faster than a commercial jet or a tornado, Kittinger reached a top speed of 988 km/h (614 mph), which is very close to the speed of sound in the stratosphere. Joe Kittinger is still the only man to fly that insanely fast without the protection of a vehicle, or even a modern space suit. His stupendous leaps during Project Excelsior weren't frivolous stunts. They showed that even with a minimum of the right equipment, a pilot could survive ejecting from his or her aircraft at the edge of space.

Q

What's the germiest thing you're likely to touch in everyday life? No one knew until The Germinator, fearless microbiologist Chuck Gerba, and his germ-busting accomplice Sheri Maxwell tracked down the filthy truth.

What is the germiest thing you touch in public?

a) bank machine

b) money

c) public toilet

d) shopping cart

The Germiest Thing You Touch in Public

What is the germiest thing you touch in public?

a) bank machine

b) money

c) public toilet

d) shopping cart

CORRECT ANSWER:

d) shopping cart

There were fewer poopy bacteria on public toilet seats than on the shopping carts they tested, which were contaminated with an average of a million bacteria each. The kiddie seats were the worst, followed by the handle. The Germinator says that if you put your groceries on the kiddie seat, "You're putting your broccoli where some kid's butt was. It's a mobile toilet!" Lucky for us, disinfectant wipes destroy most of the germs.

The Germiest Thing a Kid Touches

Q

Small children touch a zillion things a minute. They're also constantly putting their fingers into their mouths or up their noses. Little kids are blissfully unaware of the scary things they're transferring into their bodies. How about you?

What is the germiest thing a kid touches in public?

a) another kid

b) playground equipment

c) restaurant high chairs

d) shopping cart kiddie seat

The Germiest Thing a Kid Touches

What is the germiest thing a kid touches in public?

a) another kid

b) playground equipment

c) restaurant high chairs

d) shopping cart kiddie seat

CORRECT ANSWER:

b) playground equipment

Covered in all kinds of poop, bacteria, and every kind of bodily fluid, playground equipment was the most contaminated thing The Germinator tested in public. Outdoor playground equipment at fast food restaurants tended to be the worst, but the ones at schools were almost as bad. Even hours after children had touched the equipment, the germs lingered on their personal belongings and in their homes. Hand sanitizer, anyone?

Soiled Undies for Science

Q

How far would The Germinator go to track down the baddest of the bad germs? Would you believe he donated his underwear for science? To find out how many bacteria and how much poop show up in a washer load, Dr. Gerba and his students washed load after load of their own dirty undies and analyzed what came out in the wash. Let's say you washed a load of 25 pairs of underpants . . .

How much poop would you flush out? About the weight of . . .

a) a drop of mustard
b) a half teaspoon of mashed banana
c) a pinch of ground pepper
d) two American pennies

Soiled Undies for Science

A

How much poop would you flush out? About the weight of . . .

a) a drop of mustard

b) a half teaspoon of mashed banana

c) a pinch of ground pepper

d) two American pennies

CORRECT ANSWER:

b) a half teaspoon of mashed banana

The Germinator says that an average adult's soiled undies contains about a tenth of a gram of poop. Twenty-five underpants contain 2.5 grams, about the weight of a half teaspoon of mashed banana. Also found in the wash, an average of 100 million E-coli bacteria, some of which survive to invade your hands when you transfer the undies to the dryer. Then they move on to whatever you touch next. A long, hot drying cycle takes care of germy survivors in your undies, and washing your hands takes care of the rest.

THE FLAKINESS OF BEING HUMAN

Got Flakes?

Q

Dandruff. How embarrassing. Guess what? Dandruff is just the tip of the iceberg when it comes to how flaky humans really are. You're probably unaware of all the skin flakes flying off your body this very instant.

How many skin flakes do you shed in a minute?

a) 1,000
b) 10,000
c) 100,000
d) 1,000,000

Got Flakes?

How many skin flakes do you shed in a minute?

a) 1,000
b) 10,000
c) 100,000
d) 1,000,000

CORRECT ANSWER:
d) 1,000,000

You walk around in a cloud of your own liberated skin flakes. No one has actually collected and counted anyone's liberated skin particles, but some experts think that more than a million, and maybe even as many as seven million, fly off your body every minute. That adds up to anywhere from 1,440 million to more than 10,000 million skin flakes a day.

Small Flakes at Large

Your flakes crack and break away from the skin's outermost layer. Once free, the skin flakes shoot out of sleeves, pant legs, from under collars, and right through your clothes. They're obviously very small.

How small are the tiniest skin flakes? The size of a . . .

a) dust mite

b) grain of powdered sugar

c) red blood cell

d) speck of fine sand

Q

Small Flakes at Large

How small are the tiniest skin flakes? The size of a . . .

a) dust mite

b) grain of powdered sugar

c) red blood cell

d) speck of fine sand

CORRECT ANSWER:

c) red blood cell

The tiniest skin flakes are about the size of a red blood cell with a diameter of just seven thousandths of a millimetre (0.00027 in). Most of the flakes in your personal cloud are almost twice that size, about as big as white blood cells. But they're still small enough to easily fly through the gaps in the weave of the fabrics you wear.

Your Flaky Fingerprint

Q

Like your fingerprint, your personal cloud is unique, and that's what tracking dogs zero in on when following a scent. The trail gets cold when the cloud becomes too thin for the dogs to detect. Is there something that criminals fleeing from bloodhounds could do to reduce their telltale cloud of skin flakes?

What's the best way to reduce flakiness?

a) don't use scented products

b) don't use soap

c) shower

d) use skin lotion

Your Flaky Fingerprint

What's the best way to reduce flakiness?

a) don't use scented products

b) don't use soap

c) shower

d) use skin lotion

CORRECT ANSWER:

d) use skin lotion

In studies, showering with or without soap didn't reduce flakiness. Surprisingly, it increased it. Fewer flakes flew if lotion was applied after showering, but the effect only lasted for four hours. So, if you're trying to evade bloodhounds, carry a large bottle of lotion and apply it frequently. Flakiness diminishes with regular lotion use, but even bathing in lotion won't stop the shedding completely.

An average male weighing 70 kilograms (154 lb) is 63% water overall, with a total body water content of about 44 litres (46.5 quarts). A female of the same weight is around 54% water, with a total body water content of about 38 litres (40 quarts). Body water varies, but everyone's brain is equally juicy.

How watery is your brain? About as watery as . . .

a) gelatin dessert

b) raw white button mushrooms

c) raw potatoes

d) watermelon

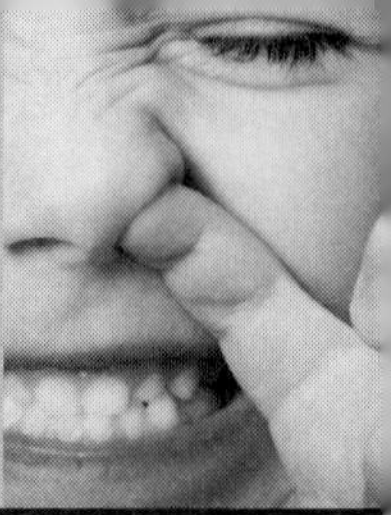

Brain Water

How watery is your brain? About as watery as . . .

a) gelatin dessert
b) raw white button mushrooms
c) raw potatoes
d) watermelon

CORRECT ANSWER:

c) raw potatoes

Raw potatoes are crunchy and, hopefully, your brain is not, but they're both about 78% water. The consistency of your brain is closer to a gelatin dessert, but the jiggling jelly is actually about 5% more watery than your brain. You wouldn't think it, but white button mushrooms and watermelon are both around 92% water.

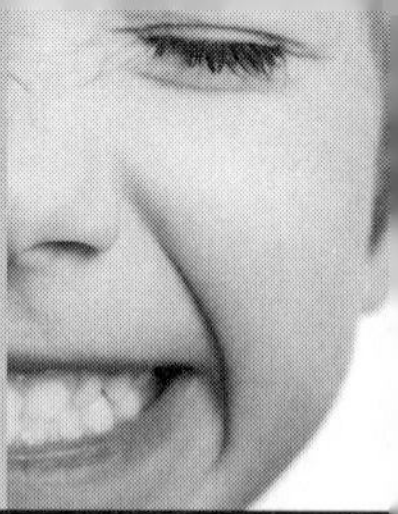

You're Leaking!

You're not aware of it, but you're leaking! Don't worry, so is everyone else. You lose water in several different ways. You exhale it from your lungs, it evaporates from your skin, and, of course, there are those eliminations that go down the toilet. The amount of water you leak doesn't vary all that much, provided you're not exercising or in an extreme environment. Ordinarily, in a day . . .

How much water do you lose? About as much as there is in . . .

a) 60 large raw eggs
b) a melon weighing 2.5 kg (5.5 lb)
c) twice the amount of your brain water
d) any of the above

You're Leaking!

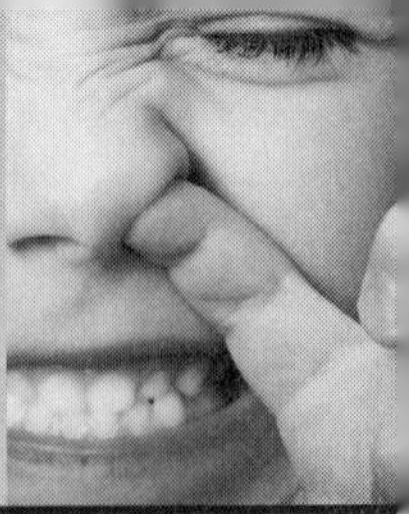

A

How much water do you lose? About as much as there is in . . .

a) 60 large raw eggs
b) a melon weighing 2.5 kg (5.5 lb)
c) twice the amount of your brain water
d) any of the above

CORRECT ANSWER:

d) any of the above

You lose about 2.4 litres (2.5 quarts) of personal fluids every day. About 0.4 litres/quarts escapes from your lungs, more in hot, dry climates or in very cold, dry places. Another 1.6 litres (1.7 quarts) or so goes down the toilet. About 400 millilitres (13.5 fl oz) evaporates from your skin at a constant rate as insensible perspiration, which means you can't feel the sweating, not that it's silly sweat.

Personal Fluid Recycling

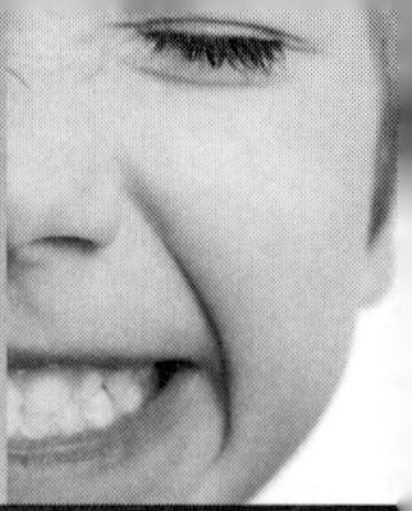

Your bodily fluids are lost to the environment through exhalation, elimination, and evaporation. But you recycle two of your bodily fluids, unless you deliberately expel some from your body. We're talking saliva and snot.

About how much saliva and snot combined do you swallow in a day?

a) 0.5 litres/quarts
b) 1 litre/quart
c) 2 litres/quarts
d) 3 litres/quarts

Personal Fluid Recycling

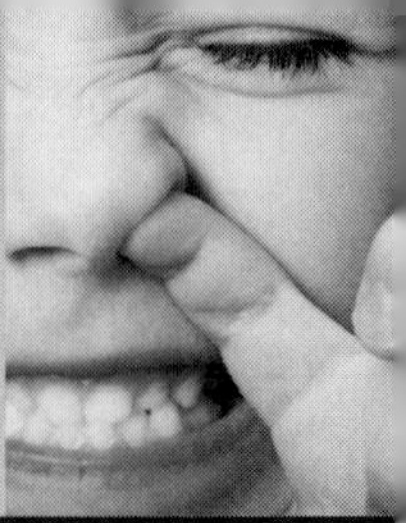

About how much saliva and snot combined do you swallow in a day?

a) 0.5 litres/quarts
b) 1 litre/quart
c) 2 litres/quarts
d) 3 litres/quarts

CORRECT ANSWER:
c) 2 litres/quarts

Every day you swallow about one litre/quart each of saliva and snot. The snot drips down the back of your throat. It's not glamorous, but it is incredibly useful. The mucus in your nose traps and hardens around foreign particles so you don't inhale them into your lungs. Some boogers float to your stomach in liquid snot. In one survey, 3% of people admitted to eating their own boogers. That's taking personal fluid recycling a bit too far, don't you think?

SNACK ATTACK: ICE CREAM

Eating Ice Cream With Your Eyes

Q

Who can resist free ice cream? That's what food psychologist Brian Wansink (mindlesseating.org) counted on when he invited students and staff from his university's nutrition department to an ice cream social. The unsuspecting guinea pigs, I mean guests, were given either a smaller or larger bowl and a smaller or larger ice cream scoop. After they'd served themselves, their ice cream was weighed. The guests with the bigger bowls and scoops took the most ice cream.

How much more ice cream did they take?

a) 27%
b) 37%
c) 47%
d) 57%

Eating Ice Cream With Your Eyes

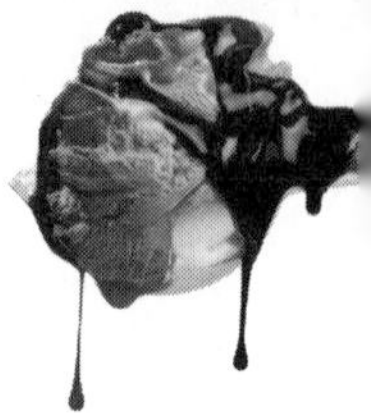

A

How much more ice cream did they take?

a) 27%

b) 37%

c) 47%

d) 57%

CORRECT ANSWER:

d) 57%

The people with bigger bowls and scoops took almost 57% more ice cream than people with smaller bowls and scoops. We eat with our eyes, and they're easily fooled. The same amount of ice cream looks like a big serving in a smaller bowl and a puny serving in a bigger one. We tend to fill the bowl to the same fullness, regardless of its size. Even the nutrition experts took 31% more ice cream if they had a bigger bowl.

Savour the Unusual Flavour

Q

Would you like some scrambled egg and bacon ice cream? The Fat Duck, chef Heston Blumenthal's restaurant in Berkshire, England, is famous for it. The adventurous chef teamed up with experimental psychologist Charles Spence to study the effect of sound on taste. They played a recording of sizzling bacon while diners tasted scrambled egg and bacon ice cream.

How did the sound influence the taste of the ice cream?

a) no influence on the taste
b) tasted more like bacon
c) tasted more like egg
d) tasted warmer

Savour the Unusual Flavour

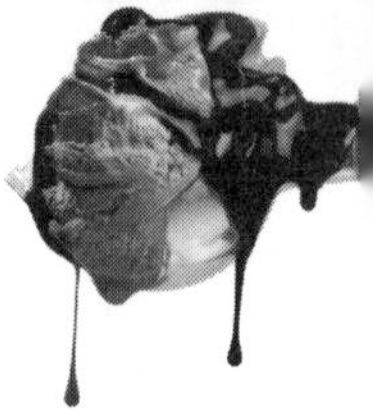

How did the sound influence the taste of the ice cream?

a) no influence on the taste

b) tasted more like bacon

c) tasted more like egg

d) tasted warmer

CORRECT ANSWER:

b) tasted more like bacon

With the sound of sizzling bacon wafting through the air, the scrambled egg and bacon ice cream tasted more like bacon. The clucking of barnyard chickens made the ice cream taste more like scrambled eggs. Ocean sounds made shellfish taste stronger and saltier. What do you think would better enhance the taste of fish and chips — the sounds of the sea or a sizzling deep fryer? If you're feeling experimental, a quick online search will lead you to some audio clips that you can listen to the next time you chow down on fish and chips.

Mmm-mm, Ice Cream!

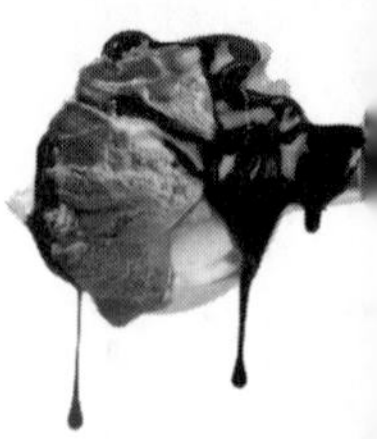

Q

Part of ice cream's appeal is the mouth feel, the luxurious sensation of it melting in your mouth. The creamy mouth feel comes from the fat in it. Premium ice creams can pack lots of fat, often 15% fat or more. Whale, dolphin, and seal milk are all about 50% fat. What would their ice cream taste like? Super creamy? Or maybe too rich? How about milk with the consistency of whipping cream?

Which animal's milk is as fatty as whipping cream?

a) dog
b) donkey
c) mouse
d) pig

Mmm-mm, Ice Cream!

Which animal's milk is as fatty as whipping cream?

a) dog

b) donkey

c) mouse

d) pig

CORRECT ANSWER:

c) mouse

Mmm-mm, mice cream! To get mouse milk for testing, German scientists custom-built a tiny, mousie milking machine. They wanted to know if mouse milk really was the most human-like, as everyone assumed. They were surprised to discover that the most human-like milk comes not from rodents, but from our best friends in the animal world, dogs. Dog milk is about twice as fatty as human milk. At almost 10% fat, dog milk could, theoretically, be used to make ice cream. If you're a really adventurous chef . . .

Q

Almost everyone has something in common: most of us find certain sounds annoying. Decades ago, a team of scientists conducted a study called "Psychoacoustics of a Chilling Sound." They asked volunteers to listen to really annoying noises and rate them.

What was rated the worst?

a) dogs barking

b) fingernails scraping on a chalkboard

c) mosquitoes buzzing

d) two pieces of Styrofoam rubbing together

The Most Annoying Sound

What was rated the worst?

a) dogs barking

b) fingernails scraping on a chalkboard

c) mosquitoes buzzing

d) two pieces of Styrofoam rubbing together

CORRECT ANSWER:

b) fingernails scraping on a chalkboard

To spare their fingernails, the researchers scraped a sound-alike metal tool across a chalkboard. Why was it rated the worst? Who knows. Even the researchers disagree. One thinks the sound is a lot like chimp warning calls, to which our primal brain might respond. Another thinks that it's the idea of scraping your nails that gives you the shivers, not the sound itself. It's not a hot research topic, so it might be a while before we know exactly why the sound of fingernails on a chalkboard sets our teeth on edge.

What Does the Earth Sound Like?

Q

The Earth is humming. Can you hear it? No? Me neither. It's an extremely deep rumble in a range of sound that we can't hear called infrasound. In musical terms, the Earth's hum is 16 octaves below a piano's middle C note. It's far too low for our ears to detect, but the seismology equipment used to monitor earthquakes and tremors picks it up. Those who study the data have an idea of what the Earth's hum would sound like if it was processed so that you could hear it. According to seismologist Hiroo Kanamori . . .

What does the Earth sound like?

a) lush, like a full orchestra playing

b) noisy, like banging on a trash can

c) Om

d) the whir of a spinning top

What Does the Earth Sound Like?

What does the Earth sound like?

a) lush, like a full orchestra playing

b) noisy, like banging on a trash can

c) Om

d) the whir of a spinning top

CORRECT ANSWER:

b) noisy, like banging on a trash can

It's not a peaceful or pretty sound. Imagine a 50-piece orchestra with none of the instruments tuned and each playing a different song. When that whole orchestra plays together, it's jarring. Maybe it's a good thing you can't hear the Earth's hum. Many experts think the sound is from thumping, twisting ocean waves, but not everyone agrees. The source of the Earth's hum is still a bit of a mystery.

Oceans of Sound

When you think of noise pollution, what comes to mind? A busy city street? An airport? A raceway? You're probably not thinking of the oceans, but there's noise pollution underwater too. Ships, industries near shorelines, dredging operations, and even marine creatures all contribute to the noise levels. In the shallow ocean . . .

What's the source of the most background noise?

a) fish farting
b) motorized water sport vehicles
c) parrotfish chomping on coral
d) shrimp snapping their claws

Oceans of Sound

What's the source of the most background noise?

a) fish farting

b) motorized water sport vehicles

c) parrotfish chomping on coral

d) shrimp snapping their claws

CORRECT ANSWER:

d) shrimp snapping their claws

Most of the background noise in the shallow ocean is caused by busy little snapping shrimp. Their claws clamp shut at an incredible 100 km/h (62 mph), so fast that an air bubble forms. When the bubble pops, it makes a sharp, crackling sound. The shrimp are plentiful and the noise is loud enough to interfere with sonar. Submarines can even avoid detection by hiding behind the sound of little shrimp snapping their claws.

Q

The anatum peregrine falcon was snatched back from the brink of extinction just in the knick of time. In the early 1970s, scientists realized that DDT, a pesticide, had wiped out most of the birds. They started captive breeding programs to replenish the wild population. Natural breeding was preferred, but some females needed artificial insemination. Can you imagine trying to get a falcon to donate sperm? How ever would you get one to do that?

How did some breeders get peregrines to donate sperm? By . . .

a) courting them while wearing a wide-brimmed hat

b) playing taped mating calls from inside realistic-looking fake females

c) relaxing them with fresh chicken meat and romantic music

d) showing them videos of other peregrines mating

Seduction, Peregrine Style

How did some breeders get peregrines to donate sperm? By . . .

a) courting them while wearing a wide-brimmed hat
b) playing taped mating calls from inside realistic-looking fake females
c) relaxing them with fresh chicken meat and romantic music
d) showing them videos of other peregrines mating

CORRECT ANSWER:

a) courting them while wearing a wide-brimmed hat

The hat trick was developed by breeders in Alberta, Canada. They discovered that some of their male falcons were really turned on by a human mimicking the peregrine courtship ritual while wearing a particular wide-brimmed hat. After sufficient chirping and bowing, the males would leap on the hat and "mate" with the brim. That's seduction, peregrine style.

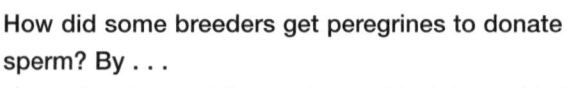

Hatching the Future

Q

How do you trick a peregrine into laying three times as many eggs as normal? Canadian breeders discovered that if they removed the eggs from the nest, the female would lay more to replace them. Using their sneaky technique, they could get up to a dozen eggs instead of the three or four a female typically lays. Soon enough, they had more eggs than peregrines to incubate them. So they recruited surrogate birds to sit in for the endangered falcons.

Which type of bird incubated the most peregrine eggs?

a) bantam chicken
b) Cooper's hawk
c) gyrfalcon
d) prairie falcon

Hatching the Future

Which type of bird incubated the most peregrine eggs?

a) bantam chicken
b) Cooper's hawk
c) gyrfalcon
d) prairie falcon

CORRECT ANSWER:

a) bantam chicken

Bantam chickens were the perfect surrogates. The little chickens were plentiful, easy to get, and perfectly willing to incubate the peregrine eggs. They had no idea they were fostering a fierce predator that would gladly have them for lunch. Some gyrfalcons and prairie falcons were also used. It took decades of dedication, hard work, and creative thinking to bring the anatum peregrine falcon back.

Sex in the City

Peregrine falcons are the poster bird for raptor recovery. When they move into urban centres, they soar to the top of the local celebrity A-list. Falcon fans even set up 24-hour webcams, the better to share the secret lives of *their* peregrines with the world. It's obvious that cities are fans of the peregrine, but why are peregrines fans of cities? If you're thinking habitat loss, you're correct, but there might be more to it than that.

Pick two things that peregrines like about cities.

a) fast food outlets

b) garbage cans

c) office towers

d) pigeons

Q

Sex in the City

Pick two things that peregrines like about cities.

a) fast food outlets

b) garbage cans

c) office towers

d) pigeons

CORRECT ANSWERS:

c) office towers and d) pigeons

In the wild, peregrines prefer high cliffs for nesting. Natural cliffs are harder to find these days, but there are lots of artificial cliffs in cities. The ledges of office buildings work as platforms for raising chicks because peregrines don't build big, elaborate nests. They mostly hunt birds, and there's no shortage of plump pigeons and other prey in urban centres. City life suits them so well, some peregrines don't even bother migrating south for the winter anymore.

GETTING A GRIP ON GECKOS

How Do Geckos Grip?

Q

Apparently, no one told geckos about gravity. They can blaze up a polished glass wall at the blistering speed of one metre (40 in) per second. When they aren't darting around, they can hang from the ceiling effortlessly. People have been trying to unravel the mystery of the gecko's incredible stickability for ages. By the turn of the 21st century, new discoveries were starting to reveal the secrets of the gecko's grip.

How do geckos grip onto surfaces? With . . .

a) gecko glue

b) invisible hairs

c) suction cups

d) sweat

How Do Geckos Grip?

How do geckos grip onto surfaces? With . . .

a) gecko glue
b) invisible hairs
c) suction cups
d) sweat

CORRECT ANSWER:

b) invisible hairs

American scientist Kellar Autumn and his team studied the Tokay gecko and discovered that there are about 6.5 million hairs, or setae, on its feet. The hairs are about 20 times thinner than a human hair and only 100 millionths of a metre (0.004 in) long. Each gecko hair has up to a thousand split ends (and you thought your hair was bad!) with tips so tiny they're invisible. It took state-of-the-art technology to discover them.

Stuck on Geckos

The hairy split ends on a Tokay gecko's feet are only 200 billionths of a metre (0.0000078 in) across. They're small enough to interact with surfaces on a molecular level. The molecules in the split ends and in the surface the gecko is on are attracted to one another with the gecko's version of The Force. Van der Waals force is a relatively weak force, but Kellar Autumn believes that the abundant split ends on a gecko's foot make the attraction strong enough to give geckos their astounding grip.

What is one split end called? A . . .

a) nanotip

b) spatula

c) spoonula

d) spreader

What is one split end called? A . . .

a) nanotip
b) spatula
c) spoonula
d) spreader

CORRECT ANSWER:

b) spatula

The word spatula comes from the Greek "spathe" meaning broad blade, which is a pretty good description of both the gecko's split ends and the common kitchen utensil. There are more than a billion spatulae on a gecko's feet. The bond they form with surfaces is so strong that a gecko hanging from smooth glass by just one hairy little toe can support eight times its own weight.

Hanging With Geckos

It takes strong arms to pry a 30-centimetre (12 in) long Tokay gecko off a surface if it doesn't want to let go. Its feet grip like Velcro-on-steroids. But when a gecko decides to move, its feet detach easily. A gecko's compact feet leave behind no residue, are self-cleaning, work underwater or in a vacuum, and on just about any surface. A million gecko foot hairs can easily fit in an area about the size of a dime (diameter 18 mm).

How much weight could a million foot hairs support?

a) 2 average women

b) an average 5-year-old child

c) an average house cat

d) an average red watermelon

Q

Hanging With Geckos

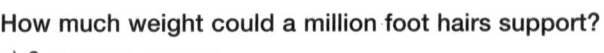

How much weight could a million foot hairs support?

a) 2 average women

b) an average 5-year-old child

c) an average house cat

d) an average red watermelon

CORRECT ANSWER:

b) an average 5-year-old child

The hairs on just one of a gecko's feet could support a five-year-old child weighing 20 kilograms (45 lb). Geckos normally use about 3% of their foot hairs at a time. But theoretically, if a gecko used all 6.5 million hairs at once, it could support the weight of two average women. Unless, of course, the gecko is on a Teflon® surface. Then all bets are off. It's the only known surface that geckos can't work up a molecular attraction with. The gecko would slide right off, just like a fried egg in your nonstick frying pan.

Q

Picture a space rock the size of an apartment building 10 storeys high and 30 metres (100 ft) across. Now imagine 200 million of them hurtling past Earth — not all at once of course. Every year, thousands of asteroids, called near-Earth asteroids or NEAS, cross or come close to Earth's orbit. They range in size from more than a kilometre across to tiny pebbles. Should you be worried? In your lifetime . . .

What are the chances of being struck by an asteroid?

a) 1 in 100 million

b) 1 in 150,000

c) 1 in 25,000

d) 1 in 3,000

The Odds of Being Struck by an Asteroid

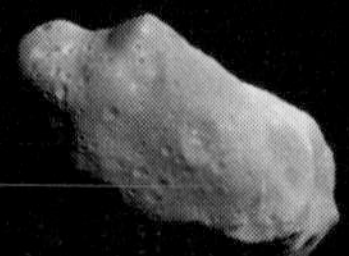

What are the chances of being struck by an asteroid?

a) 1 in 100 million

b) 1 in 150,000

c) 1 in 25,000

d) 1 in 3,000

CORRECT ANSWER:

c) 1 in 25,000

The odds of dying as a result of an asteroid strike are about the same as dying in a plane crash. However, tiny bits of asteroids bounce off you every day. You just don't notice them because they rain down as space dust. Apartment building–sized space rocks are the smallest ones with the potential to cause significant damage. Most asteroids self-destruct in the atmosphere, but once every 100 years or so a kamikaze space rock that size slams into the planet.

Slammin' Space Rocks

A space rock the size of a 10-storey apartment building exploded with the force of a three megaton bomb above Tunguska, Siberia, on June 30, 1908. The blast flattened trees in more than 2,000 square kilometres (772 sq mi) of forest. The fiery explosion lit up the sky as far away as England for several nights afterwards. That's intense. Can you imagine what a bigger asteroid the size of the Roman Colosseum, or a large sports stadium, might do?

A space rock the size of a large sports stadium could . . .

a) devastate an area the size of greater Los Angeles, California

b) generate an ocean tsunami 10 metres (33 ft) high

c) kill millions of people

d) all of the above

Slammin' Space Rocks

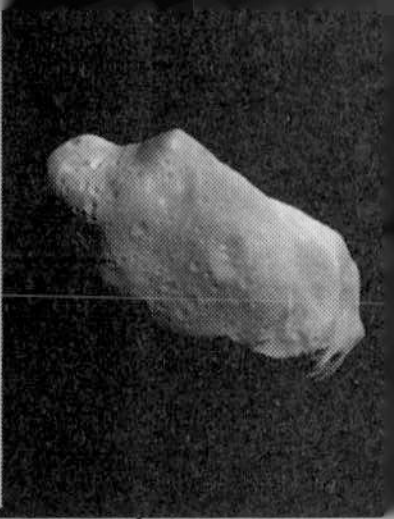

A space rock the size of a large sports stadium could . . .

a) devastate an area the size of greater Los Angeles, California
b) generate an ocean tsunami 10 metres (33 ft) high
c) kill millions of people
d) all of the above

CORRECT ANSWER:
d) all of the above

Tens of thousands of space rocks the size of a stadium, measuring 200 metres (660 ft) in diameter, cross Earth's orbit. Every two thousand years or so, a Roman Colosseum–sized rock crashes into the planet with the explosive energy of a 500 megaton bomb, minus the radiation. The damage from the impact would be on par with some of the world's worst natural disasters.

Grand Slam

About once every 100 million years, an asteroid with a diameter of at least 10 kilometres (9.3 mi) smashes into the Earth. A rock that size doomed the dinosaurs when it slammed into the Yucatán peninsula 65 million years ago. It blasted out a crater 185 kilometres (115 mi) across and almost a kilometre (3,000 ft) deep that we now know as Chicxulub.

How much energy did the impact release? About as much as a . . .

a) 25 million megaton bomb
b) 50 million megaton bomb
c) 100 million megaton bomb
d) 200 million megaton bomb

Grand Slam

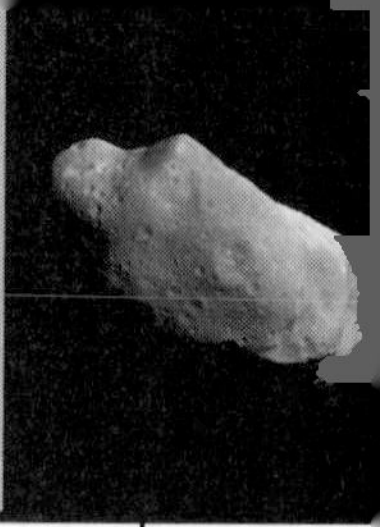

How much energy did the impact release? About as much as a . . .

a) 25 million megaton bomb
b) 50 million megaton bomb
c) 100 million megaton bomb
d) 200 million megaton bomb

CORRECT ANSWER:
c) 100 million megaton bomb

A space rock that size has the potential to annihilate an area the size of Europe, blanket Earth in a thick cloud of dust, trigger earthquakes, and set off a tsunami big enough to devastate the world's shorelines. When the Chicxulub meteor struck, the dinosaurs had already been around for about 165 million years and had survived two mass extinctions. The third time obviously wasn't the charm. If a space rock that big hits in our lifetime, we'd more than likely go the way of the dinosaurs too. To see which near-Earth objects we'll encounter today, check out spaceweather.com.

These words are choice, in more ways than one. They're interesting words that have something to do with previous questions, or questions yet to come, and you get to choose what you think they mean.

Indican:

a) beer can that changes colour when chilled
b) indicator chemical used in pregnancy test
c) radioactive tracer developed in Canada
d) salt found in sweat and urine

Bolide:

a) alternative to botox for cosmetic injections
b) fireball
c) metal pellet from iron ore smelting
d) small knot in tree roots

Euvolemia:

a) bacterial infection of the blood
b) inability to vomit
c) state of normal body fluid volume
d) supercontinent that existed on Earth 3 billion years ago

Q

Indican

CORRECT ANSWER:

d) salt found in sweat and urine

Indican is a salt excreted by your body. It's the result of bacterial activity in your guts.

Bolide

CORRECT ANSWER:

b) fireball

Bolide comes from the Greek word "bolis," meaning a missile or flash. Geologists use the word to describe any crater-forming projectile, while astronomers use it to describe an exceptionally bright fireball, particularly one that explodes.

Euvolemia

CORRECT ANSWER:

c) state of normal body fluid volume

When you replace the fluids your body loses every day, your body is in balance, a state of euvolemia.

CHOICERAMA
KELNWORDSI
EDODWHATBM
FSLICLUSOE
FNOCEBOVLL
EACACESAIO
CRANIUMRDV
TFLUFFEVEU
ASTROBLEME

What Is . . . ?

Here are some more choice words that have something to do with previous questions, or questions yet to come. What do you think they mean?

Nocebo effect:
a) effect of having no oil glands in the skin
b) gravitational pull that changes an object's orbit
c) opposite of the placebo effect
d) skin paleness from lack of sunlight during Arctic winter

Local fluff:
a) interstellar cloud through which the solar system is moving
b) lint and dust bunnies under your bed
c) office workers hired for their looks
d) plastic that washes up on shorelines

Lek:
a) area used by a group of males for courtship displays
b) encampment enclosed by woven branches and brush
c) light-emitting kiosk
d) unit of measurement for distance

Nocebo effect
CORRECT ANSWER:
c) opposite of the placebo effect

In studies, the effects of real drugs and treatments are tested against placebos, which range from pills made of inactive ingredients to sham treatments that do nothing. Yet, placebos can have the same effect as a real drug or treatment. When patients improve while receiving placebos, it's called the placebo effect. When they feel worse, it's called the nocebo effect.

Local fluff
CORRECT ANSWER:
a) interstellar cloud through which the solar system is moving

In astronomer-speak, local fluff is the nickname of the local interstellar cloud, a huge cloud roughly 30 light years across. Our solar system has been passing through it for at least 44,000 years and maybe as long as 150,000 years.

Lek
CORRECT ANSWER:
a) area used by a group of males for courtship displays

Males with small, closely packed territories gather for courtship displays in an area called a lek. The group of males is also called a lek. The males of some bird, fish, mammal, and insect species lek in large groups, the better to attract females looking to mate.

What Is . . . ?

These choice words all have something to do with previous questions, or questions yet to come. What could they possibly mean?

Astrobleme:

a) ancient astronomical instrument
b) pimple astronauts get in microgravity
c) pocket of interstellar gas
d) site of ancient meteor impact

Sulci:

a) depressions on the surface of a tooth
b) grooves in your brain
c) trenches and ditches
d) all of the above

Varve:

a) annual layer of sediment
b) saw used to cut diamonds
c) style and confidence
d) vertical carvings

What Is . . . ?

Astrobleme
CORRECT ANSWER:
d) site of ancient meteor impact

Astrobleme literally means star wound. The word comes from the Greek words *astron* (star) and *blema* (wound) and describes an ancient, eroded meteor crater, a roughly circular area of crushed, deformed bedrock. Chicxulub crater in Mexico's Yucatán peninsula is the third-largest known astrobleme on the planet.

Sulci
CORRECT ANSWER:
d) all of the above

Sulci is the plural of sulcus, which means linear groove, furrow, or slight depression. Sulci can refer to depressions on your teeth or grooves in your brain. In planetary geology, trenches and ditches found on planets and moons are called sulci.

Varve
CORRECT ANSWER:
a) annual layer of sediment

A varve is a layer of sediment or sedimentary rock laid down in one year. It's a type of rhythmite, meaning a layer that's laid down at regular intervals. Scientists "read" sediment and sedimentary rock layers to learn more about what the planet was like in the past.

Q

Some things are lost in space accidentally, and others are dumped deliberately. One astronaut was really impressed with the stuff that was routinely jettisoned from his spacecraft during his mission. He thought it was one of the most beautiful sights, maybe even the most beautiful sight, in space.

What was (possibly) the most beautiful sight in space? Jettisoned . . .

a) coolant

b) plastic garbage bags

c) rocket fuel

d) urine

The Most Beautiful Sight in Space

What was (possibly) the most beautiful sight in space? Jettisoned . . .

a) coolant

b) plastic garbage bags

c) rocket fuel

d) urine

CORRECT ANSWER:

d) urine

Apollo 9 astronaut Russell Schweickart found sunset urine dumps to be particularly spectacular. The urine flashed into millions of tiny ice crystals in the coldness of space, sort of like miniature fireworks. The golden urine crystals shooting into space were probably prettier than the 200 garbage bags dumped from the Russian space station Mir during its 15-year mission.

The Fluffiest and the Densest

Q

You probably know that Saturn is the second-largest planet in our solar system, but did you know that it's the fluffiest? The ringed planet is only about two-thirds as dense as water, which means it could, hypothetically, float like a pool toy if you could find a pool big enough to hold it. The gas giants in the solar system — Jupiter, Saturn, Neptune, and Uranus — only wish they were as dense as the terrestrial planets.

Which planet is the densest?

a) Earth

b) Mars

c) Mercury

d) Venus

The Fluffiest and the Densest

Which planet is the densest?

a) Earth
b) Mars
c) Mercury
d) Venus

CORRECT ANSWER:

a) Earth

We're number one! Earth beats out all the other planets, moons, asteroids, comets, and even the sun for the densest body. Earth, Mercury, Venus, Adrastea (one of the smallest moons), and Mars are the top five densest known bodies in the solar system.

Top 10 Densest Bodies in the Solar System

1. Earth
2. Mercury
3. Venus
4. Adrastea*
5. Mars
6. Io*
7. Moon
8. Elara*
9. Sinope*
10. Lysithea*

* Moon of Jupiter

Most Extreme Case of Global Warming

The solar system runs from deathly cold to hellishly hot, and the hottest of the hot is Venus. With the densest atmosphere of all the planets, it has the most extreme case of global warming in the solar system. At 96% carbon dioxide, Venus's atmosphere acts like an insulating blanket, trapping heat at the surface.

How hot is it at the surface of Venus? About as hot as the . . .

a) boiling point of sulphur
b) hottest hydrothermal vent on the ocean floor
c) inside an electric oven in self-clean mode
d) all of the above

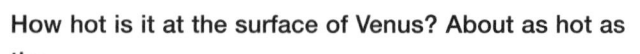

Most Extreme Case of Global Warming

How hot is it at the surface of Venus? About as hot as the . . .

a) boiling point of sulphur

b) hottest hydrothermal vent on the ocean floor

c) inside an electric oven in self-clean mode

d) all of the above

CORRECT ANSWER:

d) all of the above

Venus's surface temperature is about 460°C (860°F). The atmosphere's carbon dioxide layer is topped by swirling clouds of poisonous gas and droplets of what is, essentially, battery acid. Standing on Venus's surface would feel like being nearly a kilometre (0.6 miles) under the ocean on Earth. Venus is hot and heavy but not in a good way — at least not for Earthlings like us.

Planet George

Q

Ancient astronomers gazed up at a star-studded night sky unpolluted by artificial light. Thousands of years ago, they identified five of the six planets visible to the naked eye. They probably mistook the dim, slow-moving sixth planet for a star or comet, as did William Herschel. He spotted it in 1781 with a telescope he'd set up in his garden in Bath, England. He named the "star" for his patron, King George III of England.

Which planet was once named George?

a) Jupiter

b) Neptune

c) Saturn

d) Uranus

Planet George

Which planet was once named George?

a) Jupiter

b) Neptune

c) Saturn

d) Uranus

CORRECT ANSWER:

d) Uranus

When George proved to be a planet, astronomers clamored for a name change. It was the first planet discovered in modern times, so it was a really big deal. By the 1820s, Planet George was commonly called Uranus, which is a Latinized form of Ouranos, the name of the Greek god of the sky. The last holdout was Her Majesty's Nautical Almanac Office, which didn't switch from George to Uranus until 1850.

If Dogs Bought Their Own Toys

Q

Maybe you've heard that dogs are colour blind. Actually, dogs do see colours but not as many as a human with normal colour vision. Dogs can't discriminate between certain colours that look obviously different to us. Of the four colours listed below, three are the colours of a dog's world, and one is a colour that dogs can't see.

What colour can't dogs see?

a) blue

b) grey

c) red

d) yellow

If Dogs Bought Their Own Toys

What colour can't dogs see?

a) blue

b) grey

c) red

d) yellow

CORRECT ANSWER:

c) red

Have you ever thrown a red toy into the green grass and wondered why the dog kept running by it? It's a striking colour contrast to most of us, but dogs see both red and green as shades of grey. Their colour vision is similar to humans with red-green colour blindness. If dogs can't see red, why is it the most popular colour for dog toys? Could it be because dogs don't buy their own toys?

The World's Weirdest Lab Animal: Frankendog

Q

Dr. Vladimir Demikhov shocked the world when he revealed the world's weirdest lab animal in 1954. He'd grafted the front half of a puppy onto the neck of a large adult dog. Both dogs panted, yawned, and were hungry at the same time, but they didn't get along. The big dog kept trying to shake the puppy off, and it retaliated by biting his ear. The first Frankendog lived for just six days, but that didn't stop Dr. Demikhov from making 19 more.

Why did Dr. Demikhov make two-headed dogs? He was . . .

a) auditioning for a TV reality show
b) creating animals for freak shows
c) perfecting his surgical technique for heart transplants
d) seeing if two heads really are better than one

The World's Weirdest Lab Animal: Frankendog

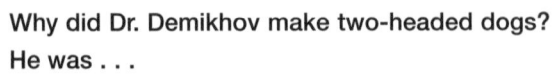

Why did Dr. Demikhov make two-headed dogs?
He was . . .

a) auditioning for a TV reality show
b) creating animals for freak shows
c) perfecting his surgical technique for heart transplants
d) seeing if two heads really are better than one

CORRECT ANSWER:

c) perfecting his surgical technique for heart transplants

None lived longer than a month, but Dr. Demikhov kept trying. He thought the problem was his surgical technique, not that the immune system of the big dog was rejecting the puppy. His dream of performing the first human heart transplant died when Dr. Christiaan Barnard beat him to it in 1967. Before the famous operation, Barnard visited Demikhov's lab a couple of times to learn from his surgical wizardry. But as far as we know, Barnard didn't make any two-headed dogs.

The Long and Short of Dog Tails

Q

The butt sniff is the official doggie handshake. But at a distance, dogs rely on the visual cues of body and tail posture. To see if short tails communicate as well as long ones, a young Canadian scientist built a fake dog — a realistic-looking, life-sized, fur-covered, black Labrador retriever with interchangeable motorized tails. The tails, one long and one short, could wag or hold still. He set the fake dog up in a popular dog park and recorded close to 500 visits by passing pooches.

Which tail state was most popular with the dogs?

a) long and wagging

b) long, upright, and still

c) short and wagging

d) short, upright, and still

The Long and Short of Dog Tails

Which tail state was most popular with the dogs?

a) long and wagging

b) long, upright, and still

c) short and wagging

d) short, upright, and still

CORRECT ANSWER:

a) long and wagging

Wagging tails signal friendliness and dogs approached the long wagging tail confidently. After sniffing where the anus should have been, they lost interest. Most stayed away if the long tail was upright and still, an unfriendly signal. Surprisingly, it didn't matter what the short tail was signalling. Wagging or not, it attracted more dogs than the long unfriendly tail (but not as many as the long friendly tail). Dogs seem to be able to read a long tail's signals more clearly than a short one's. Overall, the biggest dogs were most likely to visit the fake dog and the smallest ones to stay away.

Wolves Descending from Dogs

European wolves passed the DNA test with flying colours: their genes were 100% pure wolf. But when scientists studied wolves that had been reintroduced into the wild in North America, they were in for a surprise. They discovered that many of the wolves had a gene unique to dogs. The dog gene has such a conspicuous way of expressing itself in wolves, that even non-scientists can tell if a wolf is part dog.

How can you tell when a wolf is part dog? It . . .

a) barks when startled by noises
b) has black fur
c) jumps into the back seat of cars
d) seeks out human companionship

Wolves Descending from Dogs

How can you tell when a wolf is part dog? It . . .

a) barks when startled by noises

b) has black fur

c) jumps into the back seat of cars

d) seeks out human companionship

CORRECT ANSWER:

b) has black fur

Wolves with black fur have a genetic mutation unique to domestic dogs. The American scientists who discovered it in wolves think that dogs passed it to their wild cousins in recent history. Half the wolves in some reintroduced populations are black, and no one's sure why there are so many. In Yellowstone National Park, female grey wolves mated with black males more often than with any other colour. Is the dog gene so common because female wolves think black fur is the sexiest? Could it be that simple?

CATTITUDE!

Cats in a Strange Situation

Q

Are cats attached to their humans, or are cat lovers just imagining things? Mexican researchers explored the question with the help of 28 cats in an experiment originally conducted with human babies. In the "strange situation" experiment, babies spend time in a room with their mother and a stranger. Then, mom leaves the baby alone with the stranger and the baby's reactions are analyzed. Left alone in the room with the stranger . . .

What did the cats do? They . . .

a) acted like babies

b) played with the stranger

c) relaxed and fell asleep

d) were indifferent

Cats in a Strange Situation

What did the cats do? They . . .

a) acted like babies
b) played with the stranger
c) relaxed and fell asleep
d) were indifferent

CORRECT ANSWER:
a) acted like babies

Just like babies, when their people were present, the cats were more relaxed and likely to explore. When left with the stranger, they became more alert and much less interested in investigating. The researchers concluded that cats do form bonds with their humans. It's not news to cat people, but scientists couldn't just take someone's word for it. They're dedicated to finding proof, even if it means playing mind games with cats.

Does Absence Make a Cat's Heart Grow Fonder?

Do cats miss their owners when they're separated? How do they react to being reunited? To get to the heart of feline affection, American researchers studied the effects of quarantine on 16 cats. The owners filled out four questionnaires: at the beginning of the cat's stay in quarantine, halfway through, two weeks after quarantine, and, finally, three months after the cats left quarantine.

What were the cats like three months after quarantine?

a) more affectionate, friendly, and timid
b) more affectionate, nervous, and vocal
c) more independent, detached, and silent
d) more solitary, nervous, and timid

Does Absence Make a Cat's Heart Grow Fonder?

A

What were the cats like three months after quarantine?

a) more affectionate, friendly, and timid

b) more affectionate, nervous, and vocal

c) more independent, detached, and silent

d) more solitary, nervous, and timid

CORRECT ANSWER:

b) more affectionate, nervous, and vocal

Two weeks after quarantine, the cats were more affectionate, friendly, and timid. Three months later, they were still more affectionate but also more nervous and vocal. They spent more time with their owners than before, and they did it willingly! Cats would probably never admit it, but absence does make their hearts grow fonder. Whatever you do, don't let on that you know. Your cat would never be able to live it down.

Catnip Party

What turns the most dignified cat into a drooling, bouncy, psycho-kitty? Two things: catnip and the genes for appreciating it. Catnip sensitivity is inherited. Anywhere from 50 to 65% of domestic cats have the "catnip gene." The rest can't be expected to perform hilarious catrobatics because they're resistant, completely insensitive to catnip's charms.

What country's cat population is mostly catnip resistant?

a) Australia
b) Brazil
c) Canada
d) China

Q

Catnip Party

What country's cat population is mostly catnip resistant?

a) Australia

b) Brazil

c) Canada

d) China

CORRECT ANSWER:

a) Australia

Starting a catnip-toy business in Australia probably isn't a good idea. Strict quarantine laws led to a closed feline gene pool, and the gene for catnip sensitivity wasn't passed down. Poor kitties. They're missing out on the excitement and increased activity induced by the herb. To keep the fun and excitement fresh for responsive kitties, experts recommend no more than two catnip parties a week. Indulging too often diminishes catnip's stimulating effects.

Fat Cats in Weight Loss Programs

You can enroll a cat in a weight loss program, but you can't make it lose weight. German researchers knew that feeding cats less food in the lab led to weight loss, but it just wasn't happening at home. To figure out what was going on, they interviewed the owners of 60 normal weight and 60 overweight cats. When the owners of the fat cats were asked to describe their pet's body condition . . .

How many were in denial about how fat their cat really was?

a) 2%
b) 19%
c) 48%
d) 81%

Fat Cats in Weight Loss Programs

How many were in denial about how fat their cat really was?

a) 2%
b) 19%
c) 48%
d) 81%

CORRECT ANSWER:
d) 81%

The overwhelming majority of fat-cat owners were in denial. Only 19% admitted the obvious obesity. The rest thought that their cat was either just right, a bit big but not overweight, or on the borderline. Incredibly, 2% thought that their cat was too thin! Fat cats ate as much as they wanted, got lots of table scraps, and were rewarded with fattening treats. It answered the question of why cats in home weight loss programs didn't lose weight, but not the question of how to get the owners of the fat cats "with the program."

THE ELEPHANT LONG DISTANCE CALLING PLAN

How Do Elephants Call Long Distance?

Q

Elephants have the best phone plans. They don't have to buy a phone or sign a contract to get unlimited long distance calling. Their technology costs nothing to own or operate and is 100% energy efficient and non-polluting. They can't call their cousins overseas, but they can stay in touch with family and friends in the area.

How do elephants call long distance? By . . .

a) borrowing an eco-tourist's cell phone
b) grinding their teeth
c) rumbling
d) using their tusks as transmitters

How Do Elephants Call Long Distance?

How do elephants call long distance? By . . .

a) borrowing an eco-tourist's cell phone

b) grinding their teeth

c) rumbling

d) using their tusks as transmitters

CORRECT ANSWER:

c) rumbling

Elephants transmit their rumbles on a landline made of dirt and rocks. Their long distance calls vibrate through the ground in all directions. The extremely low-pitched rumbles are in a range that most humans can't hear, called infrasound. If you were observing an elephant making a long distance call, you might not hear it, but you'd probably feel the vibrations in your chest.

Long Distance Alarm Calls

Elephants listen to long distance calls with their feet. Their toes and trunks have special nerve endings that are sensitive to vibrations. They press them into the ground to pick up calls. The long distance calls let them stay in contact with family and friends, connect with potential mates, and send warnings about the presence of predators. As you can imagine, the predator warning call is one of the loudest.

How loud can an elephant's long distance call be? As loud as a . . .

a) gentle breeze

b) normal human conversation

c) rock concert

d) truck traffic

Long Distance Alarm Calls

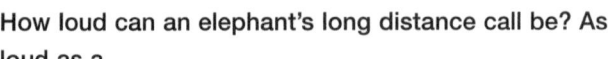

How loud can an elephant's long distance call be? As loud as a . . .

a) gentle breeze
b) normal human conversation
c) rock concert
d) truck traffic

CORRECT ANSWER:
c) rock concert

If you could hear them, elephant long distance calls would be as loud as the loudest of your four choices — a rock concert at 120 decibels! In one experiment in the wild, a recording of a matriarch making dramatic warning calls with lots of infrasound was played for various herds of elephants visiting a watering hole. They froze when they heard it and then took off in a hurry. The message was not just loud but also very clear to the elephants.

Elephant Long Distance Plan Features

Q

How would you eavesdrop on an elephant's long distance call? Scientists use sophisticated audio technology to extend their hearing into the infrasonic range. Listening in on elephant conversations has led to some amazing discoveries. Did you know that the elephant long distance plan has features you might have on your own phone?

What feature comes with the elephant long distance plan?

a) call waiting

b) caller ID

c) text messaging

d) voicemail

Elephant Long Distance Plan Features

What feature comes with the elephant long distance plan?

a) call waiting
b) caller ID
c) text messaging
d) voicemail

CORRECT ANSWER:
b) caller ID

Elephants recognize friends and family by their individual signature calls, or caller ID, and respond differently to them than to strangers. Stranger or not, bull elephants rush to answer the call of females in heat. A female's songlike mating call can go on for half an hour. By listening with their feet and trunks, the bulls can tell where and how far away she is. Hurry, hurry! She's only in heat for a few days, and it could be another four or five years before she mates again.

Time of Day Calls

You can tell when an elephant is making a long distance call by looking at its forehead. The skin vibrates as the elephant forces air through its nasal passage. The low frequencies rumble through the ground and travel much farther than they do through the air, maybe up to three times as far, depending on the terrain and time of day. The elephant long distance plan is great, but like any plan it's not perfect. For making long distance calls with the best range, certain times of the day are better than others out on the African savannah.

When is the best time to make long distance calls on the savannah?

a) mid-morning
b) noon
c) mid-afternoon
d) at night

Time of Day Calls

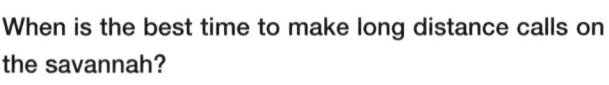

When is the best time to make long distance calls on the savannah?

a) mid-morning
b) noon
c) mid-afternoon
d) at night

CORRECT ANSWER:
d) at night

About two hours after sunset is the best time for making long distance calls on the savannah. Some scientists think that the calls travel farthest then because of nightly temperature inversions, which reflect low frequency sounds back to the ground and keep them from dissipating into the sky. An elephant's long distance calling range can increase to about 10 kilometres (6 mi) in all directions. Someone should tell the elephants because they make most of their long distance calls around 5 p.m. No one's actually asked an elephant why they don't call at night, but if you put on your elephant head, it might occur to you that calling after dark could be hazardous to your life. That's when powerful nocturnal predators, like lions, are on the prowl.

NATURAL WONDERS

Saving the World With Spider Silk

Q

Let's say you're a superhero and, in order to save the world, you have to stop a jumbo jet dead in its tracks as it flies overhead. Super confident that you will succeed, you whip out your super weapon, a lasso made out of dragline spider silk. Dragline silk is everything a spider needs it to be. It's strong, flexible, stretchy, perfect for the scaffolding of webs, and is a trusty lifeline when the spider needs to jump. But stopping a jumbo jet in mid-flight will take more than a web's worth of spider silk.

How thick is your spider silk rope? As thick as a . . .

a) fire truck hose

b) pencil

c) strand of spaghetti

d) toilet drainpipe

Saving the World With Spider Silk

How thick is your spider silk rope? As thick as a . . .

a) fire truck hose

b) pencil

c) strand of spaghetti

d) toilet drainpipe

CORRECT ANSWER:

b) pencil

Dragline silk is five times stronger than steel of the same diameter. It's also stronger and more flexible than the fibres in bulletproof vests, and it's stretchier than nylon. Nobody's actually tested whether a spider silk rope the width of a pencil would stop a jumbo jet in mid-flight, but based on lab tests dragline silk is theoretically strong enough to do it. No wonder scientists have long been trying to synthesize it.

Real Head Bangers

They'd be useless as crash test dummies. North America's largest woodpeckers slam their heads into trees at about 26 km/h (16 mph). They can jackhammer wood with their beaks 20 times a second, up to 12,000 times a day. No one has actually asked woodpeckers whether they get headaches, but scientists think that they probably don't.

Why don't woodpeckers get headaches?

a) their brains are tiny

b) their heads have shock absorbers

c) their skulls act like Styrofoam

d) all of the above

Real Head Bangers

Why don't woodpeckers get headaches?

a) their brains are tiny

b) their heads have shock absorbers

c) their skulls act like Styrofoam

d) all of the above

CORRECT ANSWER:

d) all of the above

Woodpeckers are real head bangers. Their brains are tiny and have a relatively large surface area to help absorb the blows. Their thick spongy skull bone acts like Styrofoam to cushion their brains. They also have shock absorbers in the form of muscles in their heads that contract a fraction of a second before they strike. Each strike can deliver forces as high as 1,200 Gs. To put that into perspective, if a roller coaster exerts more than 5 Gs, riders start passing out.

Your Thermal Plume

The idea that your body is surrounded by an aura dates back to ancient times. It turns out that there actually is something like an aura: it's called the human thermal plume. It surrounds your body and shoots up from the top of your head like a geyser. If you could capture your thermal plume's output for one second and find a way to inflate 23-centimetre (9 in) diameter party balloons with it . . .

How many balloons would one second of your thermal plume fill?

a) 1
b) 2
c) 3
d) 6

Your Thermal Plume

How many balloons would one second of your thermal plume fill?

a) 1

b) 2

c) 3

d) 6

CORRECT ANSWER:

d) 6

Your body heats the air close to your skin, which flows upward from the tops of your feet, up over your body. As the heated air rises, it becomes faster, thicker, and more turbulent. When it reaches your head and shoulders, your thermal plume shoots up above you, swirling at 40 litres per second. Some of the things your thermal plume launches into the world are your skin flakes, clothing fibres, cosmetic particles, and even bodily fluids.

Artificial Vanilla, Naturally

Vanilla is the world's most popular spice. Vanillin, the chemical that gives vanilla that yummy flavour and aroma, is extracted from vanilla beans. They're the seedpods of a flowering orchid, and it takes about a year to grow and cure them. Because the real thing is pricey, artificial vanilla is extracted from other things and used by the food industry as well as for non-edible products. A Japanese scientist found an interesting source of vanillin.

From what did she extract vanillin?

a) cow dung

b) fish skin

c) shiitake mushrooms

d) wasabi root

Artificial Vanilla, Naturally

From what did she extract vanillin?

a) cow dung

b) fish skin

c) shiitake mushrooms

d) wasabi root

CORRECT ANSWER:

a) cow dung

Vanillin can be synthesized from chemicals or extracted from things such as lignin, from paper mill waste. Lignin puts the crunch in crunchy plants, and there's lots of it in herbivore poop. Using cow dung, Mayu Yamamoto invented a way to extract vanillin from poop and turn the processed manure into fertilizer, all in an hour. The extracted vanillin is less than half the cost of vanillin from vanilla beans. Despite the obvious advantages, big business hasn't exactly been beating down the door to Ms. Yamamoto's lab.

Q

You can scrub your hands all day long and still find thousands of germs. No amount of washing with soap or antibacterial cleansers sterilizes them. Soap removes the bacteria temporarily, but they rebound quickly. Where could they possibly be coming from? No one knew until an American dermatologist discovered their secret hideaway with an ingenious experiment.

What did he do to his subjects?

a) coated their fingertips in tar
b) made them wear latex gloves for 48 hours
c) put superglue under their nails
d) soaked their hands in mouthwash

Germy Hands

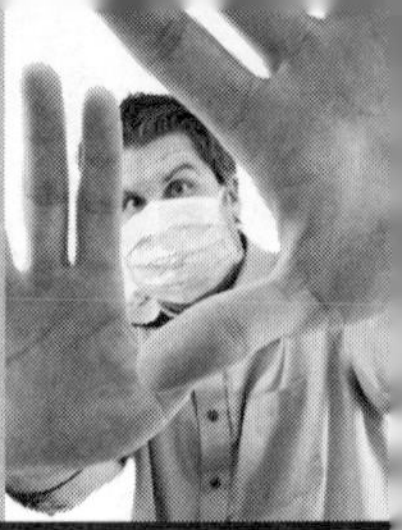

What did he do to his subjects?

a) coated their fingertips in tar
b) made them wear latex gloves for 48 hours
c) put superglue under their nails
d) soaked their hands in mouthwash

CORRECT ANSWER:

c) put superglue under their nails

James Leyden discovered hundreds of thousands, sometimes even millions, of bacteria living under people's nails. There could be more bacteria under just one fingernail than on the rest of the hand. When the moist, bacteria-friendly spaces under the nails were plugged with superglue, the number of bacteria went down each time the subjects washed their hands. Eventually, there were hardly any left. That's a good thing, right? For most of us, probably not. The vast majority of bacteria on your hands are harmless and prevent dangerous ones from getting established.

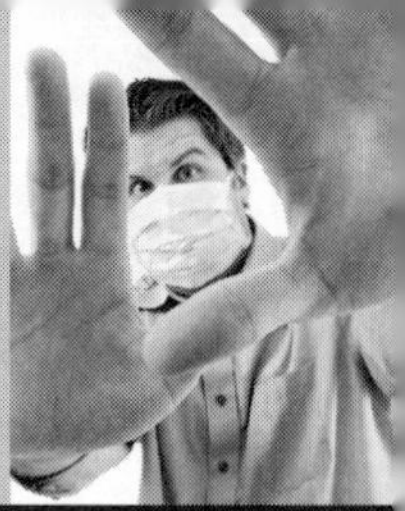

Germs on Wheels

The Germinator and his germ-busting cohort, Sheri Maxwell, turned their attention to the family car. Are cars rolling hotbeds of contamination? They tested 100 vehicles in five places across America, swabbing the steering wheel, radio knob, dashboard, door handle, seat, children's car seat, change holder, window opener, cup holder, seat belt, and food spills. All the cars were germy. Food spills were number one with the most molds and bacteria.

What was the second germiest thing in cars?

a) dashboard

b) door handle

c) radio knob

d) steering wheel

Germs on Wheels

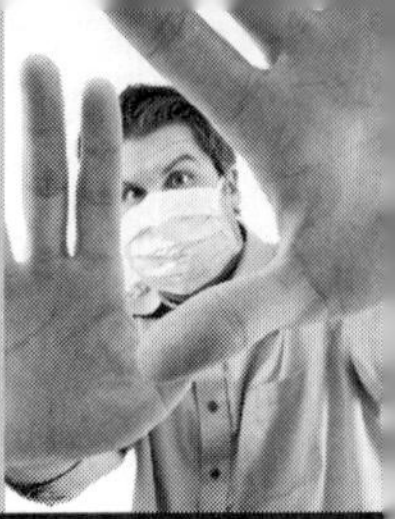

What was the second germiest thing in cars?

a) dashboard
b) door handle
c) radio knob
d) steering wheel

CORRECT ANSWER:
a) dashboard

Did you think that other parts of the car would be germier? So did the germ busters. They were surprised to find that dashboards were sometimes contaminated with as many as 800,000 bacteria in an area the size of a 10 by 10 centimetre (4 x 4 in) tile. The worst were the cars of moms in Florida, where the germs thrived in the warm, moist air. The least germy cars belonged to single men in Arizona, the state where vehicles had the fewest germs overall. Scorching, dry heat isn't ideal for germs, but even in that climate there's no such thing as a completely germ-free car.

The Top 10 Germiest Things in Cars

1. Food spills
2. Dashboard
3. Cup holder
4. Door handle
5. Seat
6. Change holder
7. Children's car seat
8. Steering wheel
9. Window opener
10. Seat belt/radio knob

The Germiest Thing in Your Hotel Room

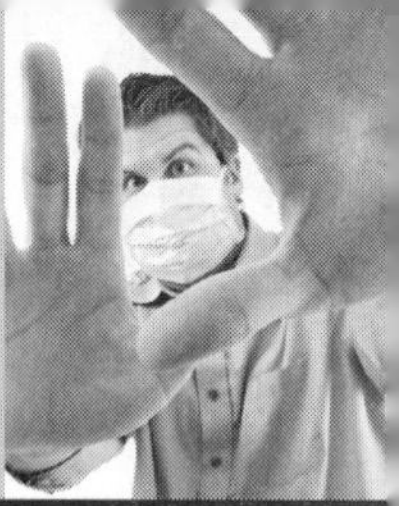

Q

On vacation at last! As soon as you arrive in your hotel room you kick off your shoes, rip off your socks, and pad around on the plush carpeting barefoot. You can't wait to go out and start enjoying your vacation, but you're a bit hungry. You check the room service menu and your favourite snack is on it, so you decide to splurge. You phone down and order. Then you flop on the bed, grab the TV remote, and flip through the channels until your nachos arrive. Bon appetit! But before you dig in, have you ever wondered . . .

What is the germiest thing in a hotel room?

a) carpet

b) phone

c) room service menu

d) TV remote

The Germiest Thing in Your Hotel Room

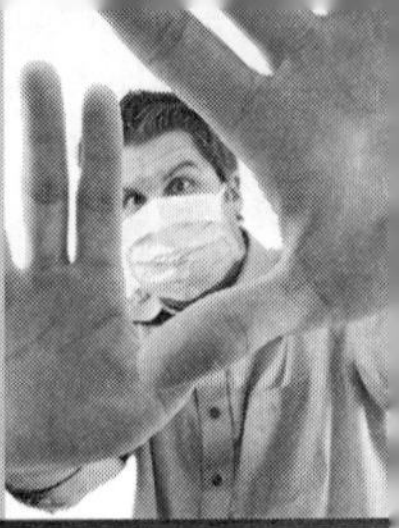

What is the germiest thing in a hotel room?

a) carpet

b) phone

c) room service menu

d) TV remote

CORRECT ANSWER:

d) TV remote

Cold and flu viruses and bad bacteria, including antibiotic-resistant ones, can survive on a remote control for up to three days. The Germinator, a.k.a. courageous microbiologist Chuck Gerba, tested hotel rooms and found fecal bacteria, such as E. coli, on remotes and phones and even on a room service menu. He also discovered every kind of bodily fluid on every bedspread, carpet, and wall. Clock radios, door handles, and light switches were often germy hot spots too. Best wash your hands before digging into those nachos, otherwise the only partying could be someone else's poopy bacteria in your guts.

Unlaundered Money

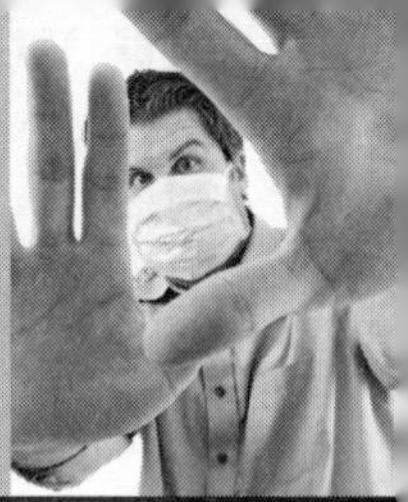

Q

Money is dirty, right? After all, it passes through many hands during its time in circulation. Who knows where it's been or who's handled it before you. How germy money is depends on how old it is, the kind of climate you live in, and how good the hygiene habits of the people who handled the money before you were. The level of contamination varies from place to place, but most of the money on the planet has some things in common.

What is the germiest?

a) coins
b) coins and bills are equally germy
c) credit or debit card
d) paper money

Unlaundered Money

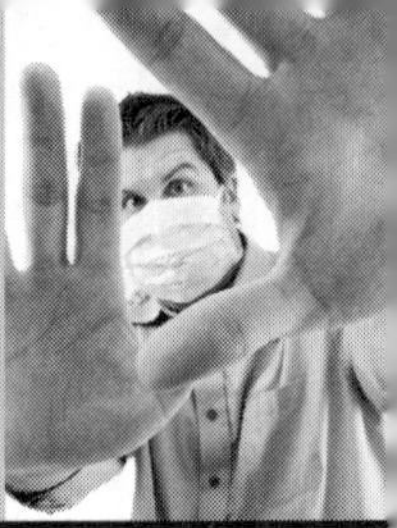

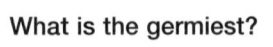

What is the germiest?

a) coins

b) coins and bills are equally germy

c) credit or debit card

d) paper money

CORRECT ANSWER:

d) paper money

Fecal bacteria showed up on bills wherever money has been studied. Some bills also carried small amounts of antibiotic-resistant bacteria and nasty viruses. Can touching contaminated bills make you ill? That depends on how many microbes are on the bill, what you do after handling the bill, and which scientist you ask. If you're paranoid about dirty money, consider moving to Japan where some bank machines steam-clean and sterilize the bills before dispensing them. Or switch to coins. The metals used to make them have antibacterial properties, so they're usually the least contaminated. Some coins even tested sterile.

Q

Tilapia, a fish that may have crossed your dinner plate, is a highly competitive species. When the time comes to mate, the males gather in a communal mating area, or lek, where they stake out the best spots they can. They want to attract lots of females and mate with as many as possible. In order to do that, the dominant males have to subjugate their rivals.

How do male tilapia assert their dominance?

a) by biting and nipping

b) by head butting

c) by urinating

d) with a colour display

King of the Fish Castle

How do male tilapia assert their dominance?

a) by biting and nipping
b) by head butting
c) by urinating
d) with a colour display

CORRECT ANSWER:
c) by urinating

Male tilapia assert their superiority with vast quantities of urine. Dominant males pee about 10 times an hour, but when there's competition, they'll go once a minute. They produce more, and much stinkier, urine than lower ranked males, and they keep at it until their opponents submit. The potent pee is more effective than you might think. Not only does it help the dominant males avoid conflict by driving off lower ranked rivals, it also attracts females looking for a studly mate.

Ants Herding Aphids

We have cows and ants have aphids, tiny insects that suck the juice out of living plants. Humans milk cows, and ants collect tiny blobs of sweet aphid excrement called honeydew. Ants love it, and they don't mind a bit of aphid meat now and then either. To ensure a steady supply of both, ants confine their herds. Humans build fences to keep their cows from wandering off. What do ants do?

How do ants control their aphid herds? By . . .

a) bullying them into submission

b) injecting them with toxic saliva

c) singing love songs to them

d) walking among them

Ants Herding Aphids

How do ants control their aphid herds? By . . .

a) bullying them into submission

b) injecting them with toxic saliva

c) singing love songs to them

d) walking among them

CORRECT ANSWER:

d) walking among them

All an ant has to do to subdue the aphids is walk among them. Their chemical footprints have a tranquilizing effect on the little bugs. When British scientists put aphids on paper that an ant had walked on, they moved much more slowly than aphids on plain paper. It's easy to see how the relationship benefits the ants, but what's in it for the aphids? It could be self-preservation. Ants fight off or kill other insects that try to poach their "cows." Maybe the aphids use the ants' chemical footprint to stay within their protection zone.

Monkey Pay-Per-View

Macaque monkeys are enough like humans to stand in for us in studies of the brain's social machinery. In a monkey version of pay-per-view, male macaques were "paid" in juice to look at imagery on a screen. It took lots of juice bribes to get them to look at some things, but there were "shows" that the monkeys loved so much they'd "pay" to see them by giving up some of their juice. Images of dominant males were a hit on monkey pay-per-view, but there was something else the monkeys gave up juice to watch.

What else did the monkeys pay to see?

a) Discovery Channel
b) female monkey bottoms
c) males ranked lower than themselves
d) themselves

Monkey Pay-Per-View

What else did the monkeys pay to see?

a) Discovery Channel

b) female monkey bottoms

c) males ranked lower than themselves

d) themselves

CORRECT ANSWER:

b) female monkey bottoms

The monkeys loved juice, yet they were willing to give some up for a glimpse of female bottoms. They paid to see dominant males but had to be bribed with extra juice to glance at lower ranked males. So what does this have to do with your brain's social machinery? The American scientists who conducted the experiment wanted to find out what motivates monkeys, and humans, to look at faces and why autistic people don't.

Monkey Cops

Dominant males are busy guys in macaque society. They're the special unit in charge of scaring off intruders, defending the troop from attackers, and defusing domestic disputes. They patrol the group like cops and settle conflicts before they can escalate. What would happen if they weren't there? Jessica Flack and her team found out when they removed several dominant males from a group of macaques at the Yerkes National Primate Research Center in the U.S.

What happened in the absence of the monkey cops?

a) all the monkeys partied and had a good time

b) nothing unusual

c) older females took the lead

d) violence escalated

Monkey Cops

What happened in the absence of the monkey cops?

a) all the monkeys partied and had a good time

b) nothing unusual

c) older females took the lead

d) violence escalated

CORRECT ANSWER:

d) violence escalated

While the monkey cops were away, cliques formed and the crime rate rose. The more unsettled the situation became, the less time the monkeys spent playing, grooming, and sitting together, and the more their social networks crumbled. It was clear that the monkey cops had enforced cooperation and stabilized the group. Without them, monkey society broke down. Many experts think that the same thing happens among humans. Surely, we're more evolved than monkeys. Aren't we?

Supersonic Ride

Q

Chuck Yeager was the first man to fly faster than the speed of sound. He broke the sound barrier in an X-1 rocket plane in 1947. The X-1 was loaded into the bomb bay of a modified B-29 Superfortress heavy bomber, which flew to 13,716 metres (45,000 ft) and launched the rocket plane. Yeager fired four powerful rocket engines and the X-1 shot to the then-incredible speed of Mach 1.06. He had a nickname for his supersonic ride.

What did Chuck Yeager call the X-1?

a) Glamorous Glennis

b) Joan Jett

c) Rockin' Millie

d) Sexy Sadie

Supersonic Ride

What did Chuck Yeager call the X-1?

a) Glamorous Glennis
b) Joan Jett
c) Rockin' Millie
d) Sexy Sadie

CORRECT ANSWER:

a) Glamorous Glennis

Yeager nicknamed every aircraft he piloted "Glamorous Glennis" for his wife. The *X-1 Glamorous Glennis* was an experimental plane that looked like a .50 calibre bullet with wings. Like a bullet, the rocket plane was designed to be stable at supersonic speeds and to withstand 18 Gs. The X-1 was the first high speed aircraft built solely for aviation research purposes, but it wasn't the last. Experimental X-planes are still being made.

Smashing Face First into a Brick Wall

Q

Dr. John Paul Stapp tore up the track at the scorching speed of 1,017 km/h (632 mph) riding the Sonic Wind I, a rocket-powered sled. He accelerated so fast he overtook a T-33 Shooting Star jet, piloted by Joe Kittinger, flying overhead. Then, the sled suddenly slammed to a stop in 1.4 seconds. The force of the deceleration was brutal. To get an idea of how extreme it was, let's compare it to being in a car that smashes into a brick wall.

How fast would your car be travelling?

a) 40 km/h (25 mph)
b) 65 km/h (40 mph)
c) 130 km/h (80 mph)
d) 190 km/h (120 mph)

Smashing Face First into a Brick Wall

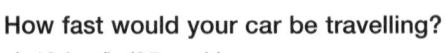

How fast would your car be travelling?

a) 40 km/h (25 mph)
b) 65 km/h (40 mph)
c) 130 km/h (80 mph)
d) 190 km/h (120 mph)

CORRECT ANSWER:
d) 190 km/h (120 mph)

Dr. John Paul Stapp served as his own crash test dummy in his rocket sled studies. He hurt his eyes during the deceleration experiment and compared the intensity of the pain to having a molar tooth extracted without anaesthetic. What motivated him to keep at it? He couldn't imagine asking anyone else to suffer the effects, according to Joe Kittinger, who thought that Dr. Stapp was the bravest man he'd ever met.

The Sonic Wind

Q

Dr. John Paul Stapp took it to the extreme on his high-speed rocket sled. To test how much force the human body could withstand, the Sonic Wind's rocket engines generated 18,144 kilograms (40,000 lb) of thrust. When the sled slammed to a halt, crushing forces of more than 40 Gs were momentarily exerted on his body. The higher the G force, the heavier your body.

How much more than normal did Dr. Stapp's body weigh?

a) 4 times
b) 10 times
c) 20 times
d) 40 times

How much more than normal did Dr. Stapp's body weigh?

a) 4 times
b) 10 times
c) 20 times
d) 40 times

CORRECT ANSWER:
d) 40 times

Dr. Stapp's normal body weight was 77 kilograms (170 lb), but for a fraction of a second he weighed 40 times as much, an incredible 3,080 kilograms (6,800 lb). That's about as heavy as an average SUV with seven people sitting in it. Data collected from dozens of increasingly brutal rocket sled rides were used in the development of safer and better helmets, aircraft seats, restraint systems, and seat belts, including the ones in your car.

Stargazer

In spirit, Project Stargazer was the father of the Hubble Space Telescope. Stargazer was designed to study the stars from the edge of space where there was no interference from the atmosphere. The craft consisted of a gondola attached to the heaviest payload balloon ever built. The project was headed by Dr. John Paul Stapp and the craft was piloted by Joe Kittinger. He took Stargazer up into the wild blue yonder in 1962.

How much could Stargazer lift to the edge of space? The weight of . . .

a) a blue whale
b) an army tank
c) the Hubble Space Telescope
d) 25 average men

Stargazer

How much could Stargazer lift to the edge of space? The weight of . . .

a) a blue whale

b) an army tank

c) the Hubble Space Telescope

d) 25 average men

CORRECT ANSWER:

d) 25 average men

The specially rigged gondola could have lifted 25 average men, but it carried just two extraordinary ones: Joe Kittinger and astronomer William C. White. Crammed into the gondola with about two tons of scientific gear, they ascended to 25,055 metres (82,200 ft) and hovered for 18.5 hours. They collected data and made observations using a telescope with a newly developed stabilization system. The mission was a success, but Project Stargazer was cancelled. The space race was on and balloons weren't in the running.

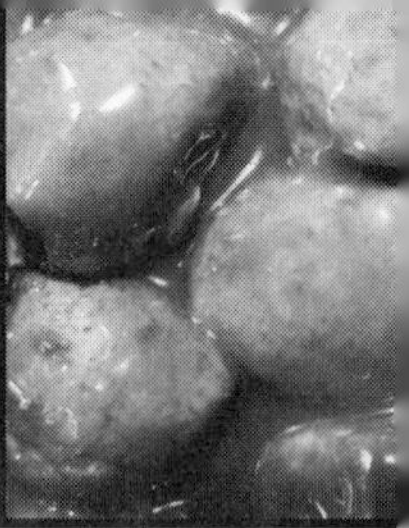

Eating in the Dark

Q

Sitting in total darkness, volunteers tasted a new strawberry yogourt and rated its yum factor. They'd been told it was being considered by the military, and since soldiers often eat in the dark, they would too. In yet another sneaky, brilliant Brian Wansink experiment, the volunteers were more in the dark than they knew. They were told they'd be eating strawberry but once the lights were out, they were given chocolate yogourt instead. Would they notice the difference?

How many rated the chocolate yogourt as having a good strawberry taste?

a) 3%
b) 20%
c) 30%
d) 60%

Eating in the Dark

How many rated the chocolate yogourt as having a good strawberry taste?

a) 3%
b) 20%
c) 30%
d) 60%

CORRECT ANSWER:
d) 60%

Strawberry and chocolate taste nothing alike, so why were so many so completely fooled? It's incredibly simple. The subjects expected strawberry, so that's what they tasted. People eat with their eyes, and since the volunteers couldn't see the yogourt's chocolate-brown colour, they tasted what they expected to taste. How many would have mistaken chocolate yogourt for strawberry if the lights had been on? Probably not nearly as many. Find out more about Wansink's eating experiments at mindlesseating.org.

A Meatball by Any Other Name Doesn't Taste the Same

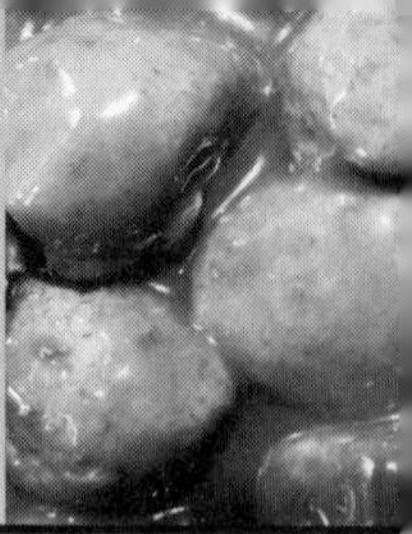

Q

Imagine you've agreed to be part of a taste test. Someone offers you "meatballs." You rate them "bland." Next, you try "spicy meatballs," find them tastier, and rate them "deeply spicy." The third meatball tastes so good, you rate it "delicious." Oh, really? Because this is a Brian Wansink experiment, all the meatballs are identical — cheap, bland, canned balls of mystery meat. The only thing that changed was how they were described to you.

Which description made the third meatball absolutely delicious?

a) imported from Italy
b) made with a gourmet blend of herbs and spices
c) old family recipe
d) traditional style

A Meatball by Any Other Name Doesn't Taste the Same

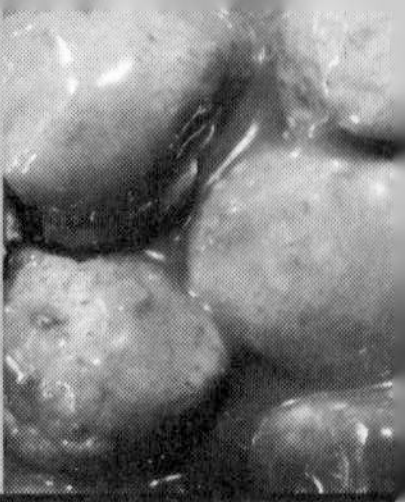

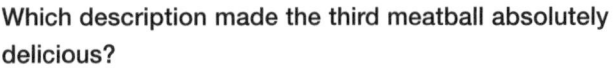

Which description made the third meatball absolutely delicious?

a) imported from Italy

b) made with a gourmet blend of herbs and spices

c) old family recipe

d) traditional style

CORRECT ANSWER:

c) old family recipe

If you thought any of the descriptions would have improved the taste of the meatballs, you're on point. So, give yourself one. In this particular experiment, the words "old family recipe" were the power words used to appeal to the tasters' most basic foodie emotions. "Family," "traditional," and "homemade" evoke nostalgic feelings about food. Place names like Italy and words like gourmet make us think the food is higher quality. Spicy, creamy, and tender are loaded with sensory appeal. Words have the power to make bland, canned meatballs taste spicy and even delicious! Whatever you do, don't tell the food industry! Oops. Too late.

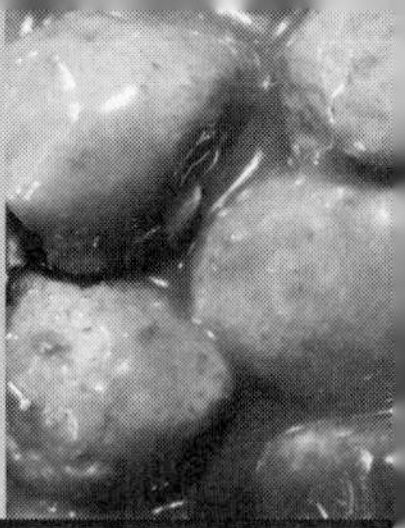

A Brownie for Your Thoughts

What's for dessert? Someone offers you a free brownie on a paper napkin and asks you to rate it. They tell you the cafeteria is thinking of adding the brownies to the menu and asks how much you'd pay for it. You think it's "nothing special" and would pay 50¢, maybe 55¢. Another batch of brownies comes out of the kitchen and your friend gets one on a white china plate. She rates it "excellent" and would be willing to pay $1.27 for it.

Why was your friend willing to pay more than twice as much for the brownie? Because it was . . .

a) bigger
b) fresher
c) served on a china plate
d) smaller

A Brownie for Your Thoughts

Why was your friend willing to pay more than twice as much for the brownie? Because it was . . .

a) bigger

b) fresher

c) served on a china plate

d) smaller

CORRECT ANSWER:

c) served on a china plate

It was a Brian Wansink experiment, so the brownies were identical. How could the same brownie go from "nothing special" to "excellent" and more than double in value? It's the magic of food psychology in the form of a nice white plate. Good presentation enhances the value of food and makes it look and taste better. That's handy to know the next time you burn the meatloaf. If you coat it in sauce or ketchup, serve it on a nice plate, and say it's a new Italian gourmet recipe you're trying, you might just get away with it.

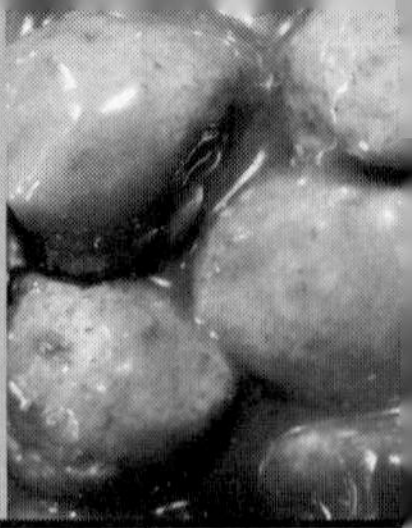

Exactly the Same, But Completely Different

A group of restaurant patrons all ate exactly the same French meal and drank the same wine. The only difference was the wine label. Because it was a Brian Wansink experiment, all the wine was the same cheap, $2-a-bottle plonk relabelled. The left side of the restaurant got "Noah's Winery — NEW from California," and the right side got "Noah's Winery — NEW from North Dakota."

Did the wine label make a difference?

a) No, it didn't affect the dining experience.

b) No, most drank the free wine.

c) No, once they'd tasted it, most didn't drink the wine.

d) Yes, it was the key to the dining experience.

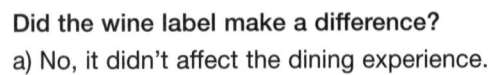

Exactly the Same, But Completely Different

Did the wine label make a difference?

a) No, it didn't affect the dining experience.

b) No, most drank the free wine.

c) No, once they'd tasted it, most didn't drink the wine.

d) Yes, it was the key to the dining experience.

CORRECT ANSWER:

d) Yes, it was the key to the dining experience.

The wine label determined how enjoyable the dining experience was. The "California" wine drinkers liked what they thought was a quality wine, and that set the tone for their entire meal. They enjoyed the food, ate 11% more, and lingered 10 minutes longer. The "North Dakota" wine drinkers weren't impressed with the wine and lowered their expectations when they saw the label. They ate less and gave the meal low marks. When Professor Wansink revealed the ruse, almost all the diners denied being influenced by the wine label. Ya think?

Q

If your brain was a car, it would be a sleek, powerful, responsive, custom-built vehicle. The media would rave about the excellence of your brain's revolutionary technology and marvel at its sophisticated design features. What would they say about its fuel efficiency? Is your brain fuel efficient or an energy guzzler? Let's say you're sitting down, working on a challenging puzzle for one hour . . .

How much fuel does your brain use?
The calories in . . .

a) 1 cup of blueberries
b) 2 chocolate chip cookies
c) 3 cups plain air-popped popcorn
d) any of the above

Feeding Your Brain

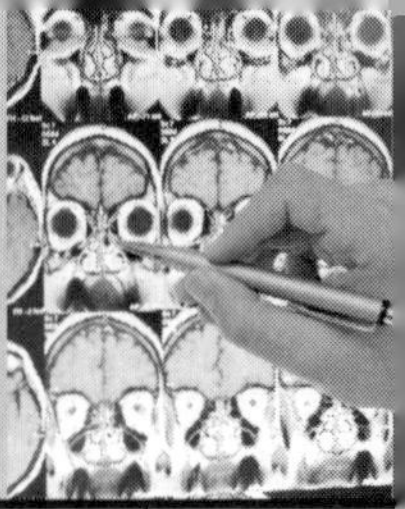

How much fuel does your brain use?
The calories in . . .

a) 1 cup of blueberries
b) 2 chocolate chip cookies
c) 3 cups plain air-popped popcorn
d) any of the above

CORRECT ANSWER:
d) any of the above

Your brain is about 2% of your weight, but it burns through about 20% of the energy from the food you eat. Your brain needs about 1.5 calories a minute, or 90 calories an hour, to work on puzzles. Where those calories come from counts. Some yummy foods help to keep your brain young, healthy, and fit. Berries, wild salmon, leafy greens, nuts and seeds, whole grains, avocado, beans, and dark chocolate are among the best superfoods for your brain.

More of What Matters

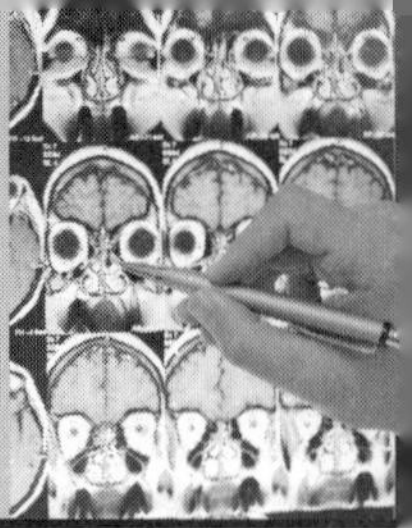

Q

Nestled inside your skull is the latest in brain evolution: your grey matter, or cerebral cortex. It's the thinking part of your brain, and in your mental prime, it packs about 25 billion nerve cells called neurons. A neuron can have up to 10,000 synapses that communicate with other cells. There could be as many as 240 trillion synapses packed into your grey matter. That's a lot of brain power crammed into your cranium! Maybe you can use it to answer this question.

How thick is the thickest part of your grey matter? As thick as . . .

a) a credit card

b) a #2 pencil

c) pita bread

d) your thumb

More of What Matters

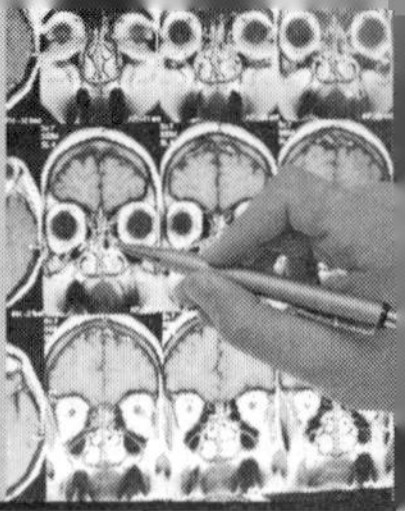

How thick is the thickest part of your grey matter?
As thick as . . .

a) a credit card
b) a #2 pencil
c) pita bread
d) your thumb

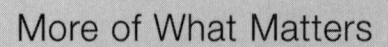

CORRECT ANSWER:
c) pita bread

At its thickest, grey matter is as thick as a flatbread that's 4.5 millimetres (0.17 in) thick. The thinnest parts of the cerebral cortex are about 1.5 millimetres (0.06 in), the thickness of two credit cards. Women have more grey matter overall, despite having smaller heads and brains. No one knows why they have more, but it could be nature's way of packing more processing power into a smaller space.

A

Hooking Up With White Matter

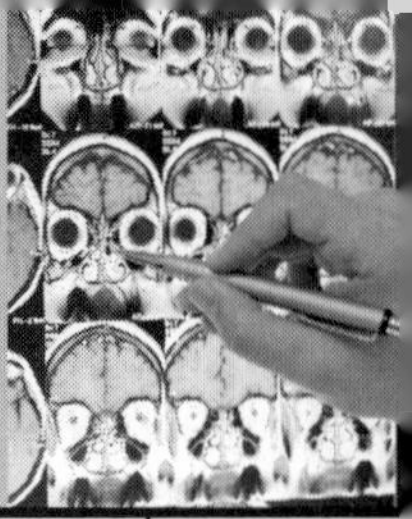

Q

Grey matter gets all the attention, but white matter is spectacular stuff too. About 60% of your brain is white matter, and it's buzzing with signals to, from, and within the brain. White matter is white because of myelin, fatty stuff that insulates the nerve cells and speeds up the conduction of impulses. What if you could take all the nerve fibres in your white matter and lay them end to end?

How long would the nerve fibre be?
Long enough to . . .

a) barely make it to the International Space Station
b) circle the Earth at the equator about four times
c) reach from Toronto, Canada, to Sydney, Australia
d) touch the moon's surface

Hooking Up With White Matter

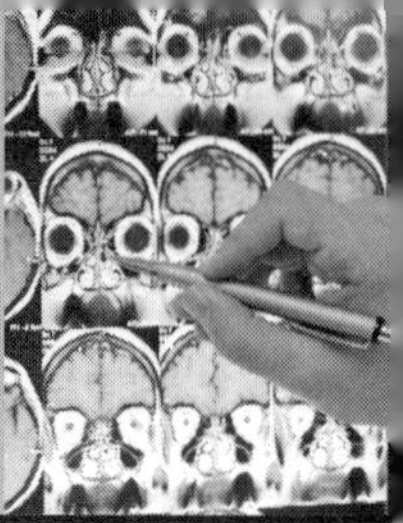

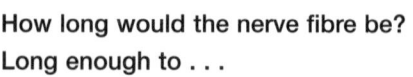

How long would the nerve fibre be?

Long enough to . . .

a) barely make it to the International Space Station
b) circle the Earth at the equator about four times
c) reach from Toronto, Canada, to Sydney, Australia
d) touch the moon's surface

CORRECT ANSWER:

b) circle the Earth at the equator about four times

Packed into an area about the size of a coconut, the brain of an average female in her mental prime has about 150,000 kilometres (93,205 mi) of nerve fibres. The average male brain in its mental prime has about 175,000 kilometres (108,740 mi) of nerve fibres. Males have more white matter and more cerebrospinal fluid. Considering that men bash their heads more often, the extra protective fluid makes perfect evolutionary sense.

Neural Speedway

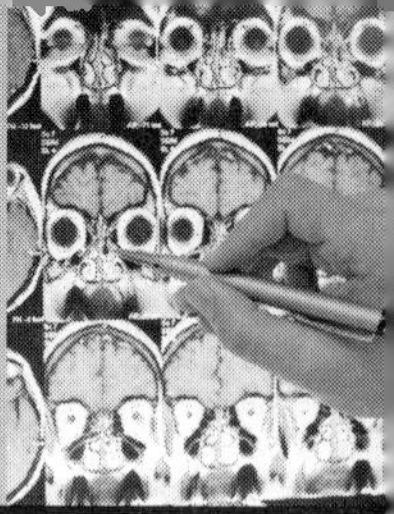

Impulses can zip along at the sizzling speed of two metres per second in your grey matter. Maybe you're thinking, "Huh? That's just over seven km/h (4.5 mph). I can run twice that fast." True, but it's lightning fast in your grey matter because of the short distances involved. The neurons are densely packed and only a couple of millimetres long. It's a different story with the nerves that run from your spine down to your leg. They can be almost a metre (3.3 ft) long and are sheathed in myelin to speed up impulses.

How fast can a myelinated nerve conduct impulses?

a) as fast as a lightning bolt

b) as fast as the world's fastest street legal car

c) at the speed of light

d) at the speed of sound

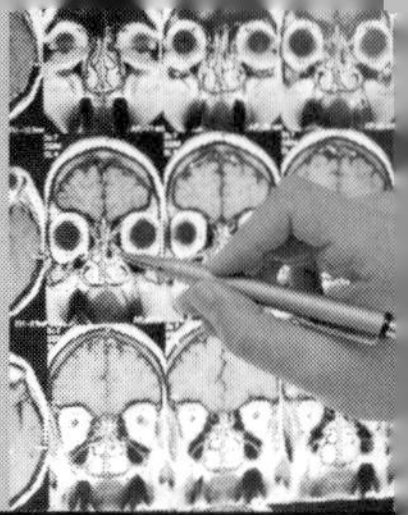

Neural Speedway

How fast can a myelinated nerve conduct impulses?

a) as fast as a lightning bolt

b) as fast as the world's fastest street legal car

c) at the speed of light

d) at the speed of sound

CORRECT ANSWER:

b) as fast as the world's fastest street legal car

The 2011 Bugatti Veyron, with an average top speed of 431 km/h (268 mph), is the world's fastest street legal car, as of this writing. A heavily myelinated nerve is comparable, conducting impulses at the same speed. It takes three or four times longer to blink than for an impulse to race from your brain to your foot. Most of the nerves that run from your brain to your muscles and organs are coated in the white stuff for speed, myelin.

Q

Mice are enough like us to stand in for us in experiments and studies in laboratories. Even though our physiology is similar, our behaviour is worlds apart. OK, maybe not worlds apart, but when was the last time you built a cozy little nest in the corner of your bedroom? Here's another example: in one study, males were exposed to the scent of a female mouse's urine.

What did the male mice do?

a) butted heads like goats

b) howled like wolves

c) sang like birds

d) urinated on their paws like capuchin monkeys

The Alluring Scent of Eau de Pee

What did the male mice do?

a) butted heads like goats

b) howled like wolves

c) sang like birds

d) urinated on their paws like capuchin monkeys

CORRECT ANSWER:

c) sang like birds

Researchers knew that male mice vocalized when they caught a whiff of female urine because they could see the rodents' little mouths moving. But they couldn't hear them because the calls are in the ultrasonic range, too high-pitched for our ears to detect. When they recorded the calls and processed them to be audible to humans, the scientists were amazed to discover that male mice sing complex songs similar to birdsong. Even more incredible, no two mousie love songs were exactly alike.

Eating Like a Stressed Rat

When it comes to the rat race, lab rats are a lot like us. They react to stress by producing stress hormones and, just like us, they develop health problems if the stress is ongoing. To keep us healthy, scientists are looking for ways to break chronic stress cycles. In one experiment rats were subjected to prolonged stress. Twenty-four hours later, they were fed something that switched off their chronic stress cycles.

What puts the brakes on the rats' chronic stress cycles?

a) drinks with artificial sweeteners
b) leafy greens and vegetables
c) sugar and fat
d) water and skim milk

Eating Like a Stressed Rat

What puts the brakes on the rats' chronic stress cycles?

a) drinks with artificial sweeteners
b) leafy greens and vegetables
c) sugar and fat
d) water and skim milk

CORRECT ANSWER:
c) sugar and fat

When the tense rats got a hit of sweet, rich, fattening foods, it reduced the levels of stress hormones and broke the cycle. Sugar affects the same brain pathways as addictive drugs and prompts a release of feel-good hormones. There's no reward pathway in your brain for artificial sweeteners. Your gut doesn't recognize the synthetic molecules, which means there's no metabolic feedback and no stress relief. If you instinctively reach for sweet or fattening foods in times of stress, now you know why. You're just trying to make yourself feel better.

Of Mice and Humans

Q

Heroic lab mice have made many medical breakthroughs possible, including organ transplants, which were first practiced on them. Mice are also used to test the effects of just about anything you can ingest, inhale, be coated in, or exposed to. They've been with us, mostly ruining our food supplies, for about 10,000 years, ever since we started storing food. Some experts think we domesticated cats in an effort to control mice. Did you know that? What else do you know about mice?

Which statement is *not* true?

a) Dominant male mice have harems of females.
b) Male mice have nipples.
c) Mice don't vomit.
d) Mice eat their own droppings.

Of Mice and Humans

Which statement is *not* true?

a) Dominant male mice have harems of females.

b) Male mice have nipples.

c) Mice don't vomit.

d) Mice eat their own droppings.

CORRECT ANSWER:

b) Male mice have nipples.

Male mice do *not* have nipples. Nobody's sure exactly why human males have them, but we do know why male mice don't. Their nipple development is suppressed at the fetal stage, which doesn't happen in human males. Does the thought of eating your own droppings make you want to vomit? Mice couldn't vomit if they tried. They're not built for it. Like most animals that recycle their own poo, mice do it to recover the nutrients they didn't absorb the first time around. You have to admit, that's an excellent survival strategy — for a mouse!

The Most Famous Lab Rat

How would you teach a pigeon to bowl? Behaviourist B.F. Skinner used positive reinforcement to shape the behaviour of a pigeon he'd caught on the windowsill of his lab. He trained it by gradually modifying its behaviour with treats. Soon, the bird was doing exactly what he wanted it to do, using its beak to roll a wooden ball down a miniature alley toward a set of toy pins. Years before he taught the pigeon to bowl, Skinner conducted an experiment that made a celebrity of his favourite lab rat, Pliny the Elder.

What did B.F. Skinner train Pliny to do?

a) drive a car

b) operate a vending machine

c) play video games

d) write a novel

The Most Famous Lab Rat

What did B.F. Skinner train Pliny to do?

a) drive a car
b) operate a vending machine
c) play video games
d) write a novel

CORRECT ANSWER:
b) operate a vending machine

When he read about chimps earning poker chips to trade for treats, Dr. Skinner wanted to show that less advanced animals could do it too. He trained his lab rat to operate a vending machine of sorts. First, Pliny released his "coin" (glass marble) from a rack by pulling a chain. Next, he picked up the marble and walked across the cage on his hind legs to a tube sticking out of the floor. He dropped the marble into the tube, and the vending machine dispensed a treat. His spending habit was featured in a photo spread in *Life* magazine in 1937, which makes Pliny the Elder the most famous lab rat of all time.

Q

The world's oceans are turning into garbage dumps. Eighty percent of the trash in the ocean comes from land and the rest is jettisoned by ships or spilled from shipping containers. The garbage collects in the middle of huge circular ocean currents called gyres. There are 11 major gyres in the world's oceans. The North Pacific Gyre is home to the Great Pacific Garbage Patch, where an estimated 100 million tons of junk has accumulated.

How much of the trash in the Great Pacific Garbage Patch is plastic?

a) 60%
b) 70%
c) 80%
d) 90%

How much of the trash in the Great Pacific Garbage Patch is plastic?

a) 60%
b) 70%
c) 80%
d) 90%

CORRECT ANSWER:
d) 90%

Most of the trash is plastic. As many as 750,000 bits per square kilometre (1.9 million per sq mi) swirl in a vortex the size of a small continent. You can't walk on it because it's not solid. It's more like plastic soup. The South Pacific Gyre is four times as big and scientists fear that it's even more polluted than the Great Pacific Garbage Patch. No one's sure how much plastic flows into the world's oceans every day. Some experts think it's about six metric tonnes (6.6 tons) a day, while others think that estimate is much too low.

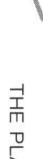

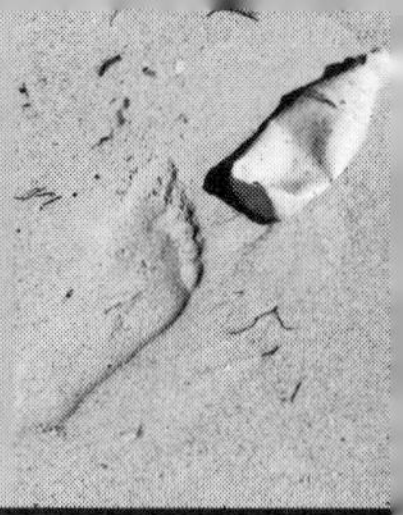

The Incredible Journey of the Yellow Duckies

Q

Every year thousands of shipping containers accidentally spill their cargo into the ocean. A few of the countless objects lost in and around the North Pacific Gyre are 80,000 Nike athletic shoes, 34,000 hockey gloves, and 29,000 plastic bath toys, including yellow duckies. To explore the world's ocean currents, Dr. Curtis Ebbesmeyer (flotsametrics.com) and Jim Ingraham track lost objects like the bath toys that began their incredible journey in 1992.

Where have the yellow duckies travelled?

a) from Alaska to Japan and back again

b) to the Arctic Ocean

c) to the North Atlantic

d) all of the above

The Incredible Journey of the Yellow Duckies

Where have the yellow duckies travelled?

a) from Alaska to Japan and back again

b) to the Arctic Ocean

c) to the North Atlantic

d) all of the above

CORRECT ANSWER:

d) all of the above

They discovered that some bath toys circled clockwise from Alaska to Japan and back while others drifted north to the Arctic Ocean. Incredibly, the bath toys crossed the Arctic Ocean by drifting 1.6 kilometres (1 mi) a day. They floated into the North Atlantic about eight years after they'd been lost at sea. The scientists continue to track the aging yellow duckies, blue turtles, green frogs, and red beavers. Most of them are still cruising the oceans or circling the world's biggest garbage dumps.

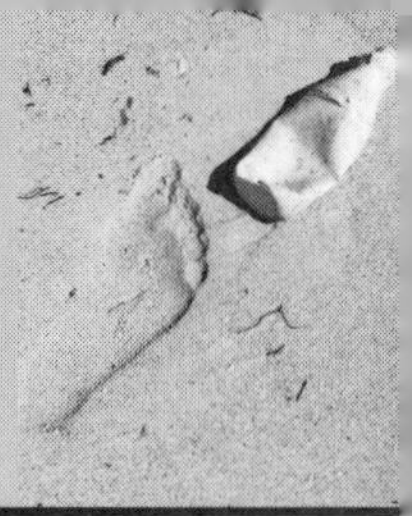

Old Plastic Never Dies

Old plastic never dies. It just gets smaller and smaller. That's what American researchers found when they went to the Great Pacific Garbage Patch in search of plankton. The tiny creatures are at the bottom of the food chain, which makes them critically important for the survival of marine life. The researchers found plankton, as well as plankton-sized bits of plastic.

How much plastic was there compared to plankton?

a) half as much

b) as much

c) three times as much

d) six times as much

Old Plastic Never Dies

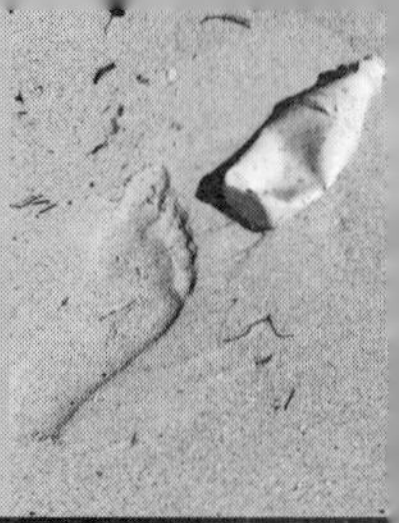

How much plastic was there compared to plankton?

a) half as much

b) as much

c) three times as much

d) six times as much

CORRECT ANSWER:

d) six times as much

The researchers were shocked to find six times more plankton-sized bits of plastic than plankton in their samples. Plastic can be up to a million times more contaminated with toxins than the surrounding water, and fish can't tell it from food. Plastic eventually breaks down into a powder fine enough for plankton to ingest. Are they eating it, and is it poisoning them? How would plastic-eating plankton affect the food chain? That's what scientists are trying to find out.

The Plastic Food Chain

There's a plastics museum in the ocean. Unless it's landed on a beach, been buried in sediment, or eaten, every piece of plastic that's gone into the ocean is still out there. Sixty years after a World War II seaplane was shot down, plastic debris from it was found in the stomach of an albatross. It's not unusual to find plastic in the stomachs of birds. A British marine biologist who studied dead sea birds along North Sea coastlines was surprised by what he found. On average . . .

How many bits of plastic were in each sea bird's stomach?

a) 44
b) 33
c) 22
d) 11

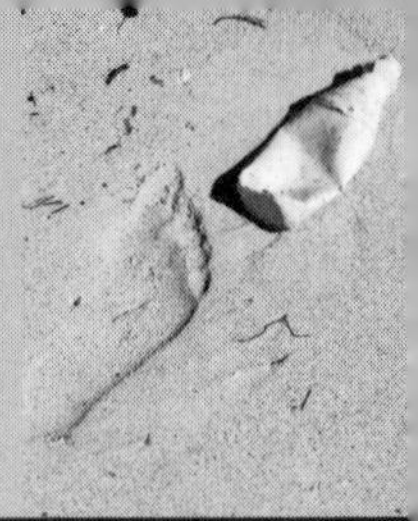

How many bits of plastic were in each sea bird's stomach?

a) 44
b) 33
c) 22
d) 11

CORRECT ANSWER:
a) 44

It was the worst case Dr. Richard Thompson had ever seen. Ninety-five percent of the dead birds had eaten large amounts of plastic. Imagine having about two kilos (almost 5 lb) of plastic in your stomach. That's the human equivalent of what the birds ate, and it probably killed them by damaging or blocking their digestive tracts. Plastic also chokes, traps, maims, and kills marine creatures. Over 30 years ago, a scientist estimated that every year more than a million seabirds and over 100,000 mammals and sea turtles were killed by plastic. Can you imagine what the real numbers are today?

Q

Musicians are trained listeners. They're better at perceiving and remembering sounds than non-musicians are. Playing in a band or orchestra, they can pick out a single instrument from among all the other competing sounds. They notice tiny differences in pitch and can tell when something is even slightly out of tune. So, they must have better ears, right? Musicians are notorious for avoiding hearing tests, but researchers managed to find a few willing to get their ears checked in the name of science.

How did the musicians' ears rate?

a) way above average

b) above average

c) average

d) below average

How Well Do Musicians Hear?

How did the musicians' ears rate?

a) way above average

b) above average

c) average

d) below average

CORRECT ANSWER:

c) average

Musicians' ears are just like everyone else's, but their brains are different. Musicians' brains were monitored while they listened to tones. Their auditory cortex, which processes sounds, reacted twice as much as a non-musician's. Asked to pick out a single voice buried in multiple noisy conversations, the musicians aced the test. Their brains could isolate a voice in the same way they could isolate an instrument. Their ears may be ordinary, but musicians' hearing is anything but.

Too Loud, Too Long

Q

Musicians and factory workers have at least one thing in common. They're exposed to loud sounds for long periods of time and develop hearing loss as a result. A German study tested musicians and factory workers with comparable levels of hearing loss to find out how they perceived music. When they were almost completely deaf . . .

Who could detect slightly off-key harmonies?

a) both musicians and factory workers

b) factory workers

c) musicians

d) neither musicians nor factory workers

Who could detect slightly off-key harmonies?

a) both musicians and factory workers
b) factory workers
c) musicians
d) neither musicians nor factory workers

CORRECT ANSWER:
c) musicians

The musicians could tell when harmonies were slightly off-key even when their ears were totally shot. They were listening with their brains, which were trained to discern pitch, timing, and tone quality — the very things that identify a voice or instrument. Factory workers with the same level of hearing loss couldn't tell the difference.

The Speed of Music

If you've ever listened to music while exercising, you know that it makes a difference. Curious about how the tempo, or speed, of music affects a workout, British psychologists recorded six popular songs. They then made a slightly faster version and a slightly slower version without altering the music's pitch. Student volunteers exercised to a different version once a week for three weeks. They didn't know about the tempo differences. When they rated the recordings . . .

Q

Which version of the music got the highest rating?

a) faster

b) normal

c) slower

d) the tempo made no difference

The Speed of Music

Which version of the music got the highest rating?

a) faster

b) normal

c) slower

d) the tempo made no difference

CORRECT ANSWER:

a) faster

The slightly faster version got the best response. It motivated the subjects to endure more discomfort, and enjoy doing it! Whether we're aware of it or not, we pace ourselves to whatever tempo is playing, so picking the tempo that suits your workout is key. The music you love makes exercise more enjoyable, and motivational lyrics can inspire you to endure the toughest parts of your routine.

How Do You Hear Silent Music?

Q

At a rock concert your ears are full to bursting with the sound of amplified music, but there's even more to the dense soundscape than what you can hear. Just under the throbbing bass sound are tones in the infrasonic range that are too low to hear. A rock concert wouldn't be the same without the silent thunder under the bass. If not with your ears, how are you hearing it?

With what do you hear silent music?

a) gut
b) hair
c) intuition
d) ribcage

How Do You Hear Silent Music?

With what do you hear silent music?

a) gut

b) hair

c) intuition

d) ribcage

CORRECT ANSWER:

d) ribcage

The sensation of bass comes from certain infrasonic frequencies that rattle your ribcage and the inside of your head. In an experiment called Soundless Music, infrasound was generated during live piano concerts. When it was playing, audience members felt 22% more strange feelings than when it wasn't. They reported a feeling of compression around the head and neck, a strange blend of tranquility and unease, like being in a jet before it takes off, and . . . an intense feeling of paying to be in an experiment. Obviously, not everyone got the same vibe from the soundless music.

PSYCH! SEEING AND BELIEVING

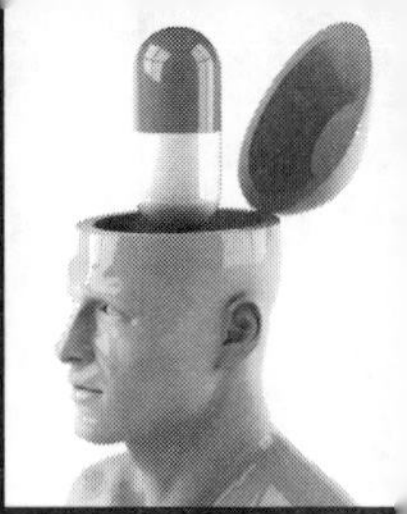

You Can't Always Get What You See

Q

You're walking along on a busy street when a young man stops you and politely asks for directions. As you start to speak, a large door passes between you. While he's out of view, he trades places with one of the two men carrying the door. After the interruption, you continue to give directions. Do you notice that you're talking to a different person? It's not a candid comedy show. It's a classic Harvard University experiment that involved 15 unsuspecting subjects . . .

How many did not notice the switch?

a) 1
b) 5
c) 8
d) 12

You Can't Always Get What You See

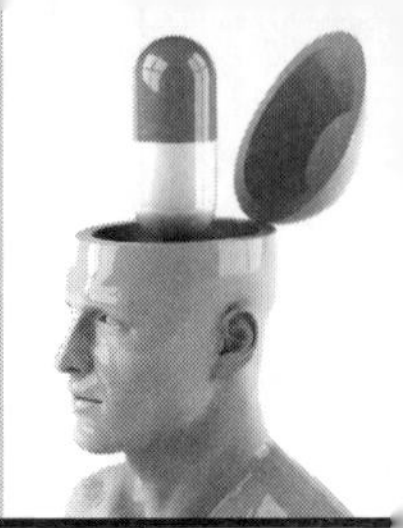

How many did not notice the switch?

a) 1

b) 5

c) 8

d) 12

CORRECT ANSWER:

c) 8

More than half experienced an episode of change blindness. If your attention is focused on one thing, like giving directions, you tend not to notice other things. A distraction, interruption, or something unexpected can blind you to the obvious, such as a different person standing in front of you. Your eyes definitely see it, but your brain doesn't register the change. It was a funny experiment, but change blindness has a serious side. It's a contributing factor in about nine out of 10 traffic accidents.

Have you ever experienced change blindness? You can check how susceptible you really are with an online test at tiny.cc/AttentionBlindness.

A Battle of the Senses

Which sense dominates your awareness? That's what a group of psychologists wanted to know. In a battle of the senses, would seeing or hearing triumph? To find out, the researchers set up the scientific battlefield: a computer screen that showed a single flash accompanied by either one or two short beeps. Volunteers were asked what they'd seen and heard.

Q

What did they sense when one flash and two beeps were played?

a) one flash and one beep
b) one flash and two beeps
c) two flashes and one beep
d) two flashes and two beeps

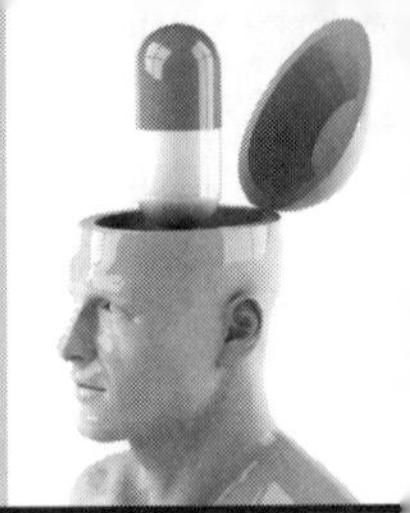

A Battle of the Senses

A

What did they sense when one flash and two beeps were played?

a) one flash and one beep
b) one flash and two beeps
c) two flashes and one beep
d) two flashes and two beeps

CORRECT ANSWER:

d) two flashes and two beeps

As long as the single flash and two beeps were close together, the subjects saw two flashes. It's not what anyone expected. The experiment (tiny.cc/FlashBeeps) cast serious doubt on the common assumption that sight always trumps sound. Perhaps it shines a headlight on the trouble with talking on cell phones, even hands-free models, while driving. The phone call and the road ahead compete for your attention, which diminishes your perception and reaction time. If anything unexpected happens, you're as accident-prone as someone who's had two or three drinks.

Cheap vs. Expensive Fake Drugs

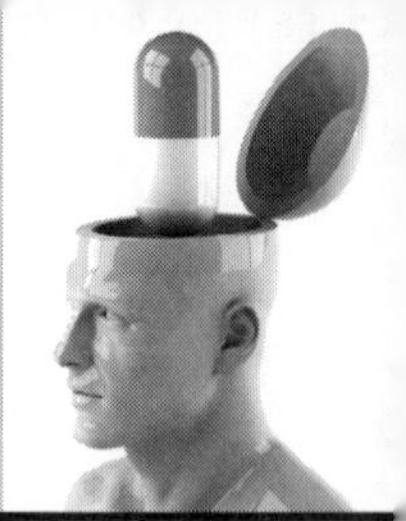

Q

Placebos are fake "drugs" made out of something harmless, like sugar. Incredibly, they can work like real drugs in some people, some of the time. When a fake drug works it's called the placebo effect. In one experiment, two groups of volunteers were told that they were getting a new kind of painkiller. One group was told that the pills cost $2.50 each and the other group that the "drug" was marked down to ten cents a pill. They were all given exactly the same placebo.

How did the placebo painkillers perform?

a) neither was effective

b) the cheap pill was more effective

c) the expensive pill was more effective

d) they were both equally effective

Cheap vs. Expensive Fake Drugs

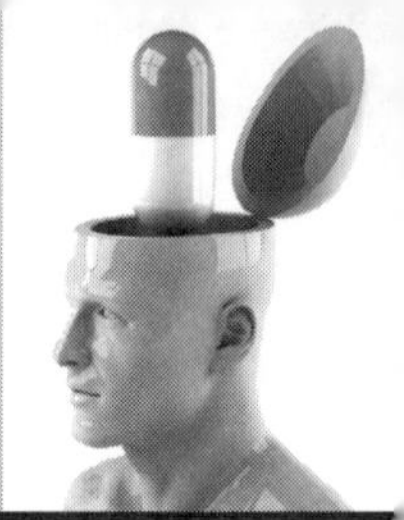

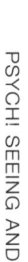

How did the placebo painkillers perform?

a) neither was effective

b) the cheap pill was more effective

c) the expensive pill was more effective

d) they were both equally effective

CORRECT ANSWER:

c) the expensive pill was more effective

Those taking the expensive placebo experienced an 85% reduction in pain compared to 61% for those taking the cheap pill. Why? People expect a more expensive drug to perform better. Once you've decided how effective a treatment will be, your brain makes it so. That's both the simple secret and the deepest mystery of how placebos work.

Placebo Enhancers

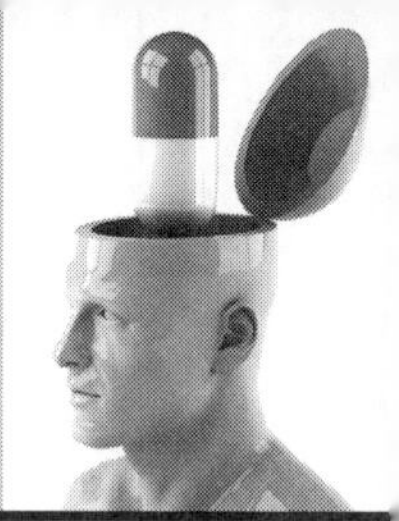

A placebo's colour can enhance its effects. Hot coloured placebos are stimulating, while cool coloured ones are the opposite. Capsules are more effective than pills, but the optimum size depends on your culture. What about when the placebo is not a drug, but a treatment? In a drug trial for irritable bowel syndrome, sham acupuncture treatments from sympathetic practitioners were tested against a real drug.

How did the placebo treatment and the drug compare?

a) both were equally effective
b) the drug was more effective
c) the placebo treatment was more effective
d) neither worked

Placebo Enhancers

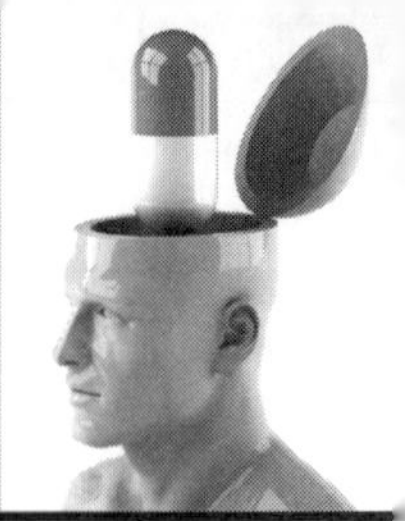

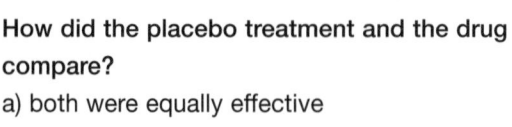

How did the placebo treatment and the drug compare?

a) both were equally effective
b) the drug was more effective
c) the placebo treatment was more effective
d) neither worked

CORRECT ANSWER:

a) both were equally effective

About 60% of the subjects got "adequate relief" from either the treatment or the drug. The sham acupuncture was performed with fake needles on non-acupuncture points, but it worked as well as the drug because the practitioner was sympathetic. A caring doctor can be a living, breathing placebo effect, and it doesn't matter what colour, size, or shape he or she is. Dr. Placebo's active ingredients — warmth, confidence, and attentiveness — can make any treatments, fake or real, more effective.

GERMyNATION: ON THE JOB

The Germiest Job

Q

Who has the germiest job? To get the real dirt on the workplace, The Germinator went on the job with bankers, accountants, consultants, doctors, lawyers, school teachers, publicists, and radio and television personnel. Microbiologist Dr. Chuck Gerba and his intrepid team tested ordinary things that people use at work. When the levels of contamination were compared, one workplace emerged as the germiest by far.

Who had the germiest job?

a) accountants

b) bankers

c) doctors

d) teachers

The Germiest Job

Who had the germiest job?

a) accountants
b) bankers
c) doctors
d) teachers

CORRECT ANSWER:

d) teachers

School teachers had up to 20 times as many bacteria in their workspace as most of the other professions. Why so many germs? Well, schools are full of children who are excellent at transporting germs. They're probably the ones transferring bacteria onto the surfaces around a teacher's stuff. Accountants and bankers had the second and third germiest jobs.

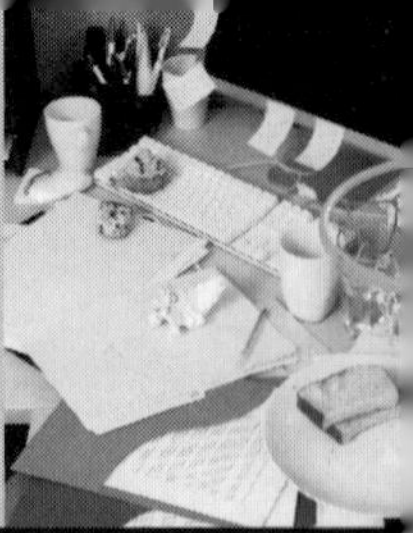

The Least Germy Job

To answer the burning questions of who has the germiest and the least germy job, Dr. Chuck Gerba, a.k.a. The Germinator, and his team tested common objects in the workplace. They swabbed phones, desks, computer keyboards, mouse, doorknobs, and pens to capture samples of the bacteria lurking on the surfaces. They discovered that one profession had almost 20 times fewer bacteria than the germiest job.

Who had the least germy job?

a) consultants
b) lawyers
c) publicists
d) television producers

The Least Germy Job

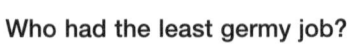

Who had the least germy job?

a) consultants
b) lawyers
c) publicists
d) television producers

CORRECT ANSWER:

b) lawyers

Overall, lawyers had the least germy work spaces. Their desks had a mere 2% of the germs found on the most contaminated desks. Why were their desks so clean? Most of the bacteria on desks come from food eaten at them. Maybe the lawyers in the study ate out more than the other professionals. Publicists and consultants, who had the second and third least germy jobs, often eat out when meeting clients too. Eating out may be a way to reduce the number of germs on your desk, but if you don't make as much money as a lawyer, disinfectant wipes are an economical alternative.

The Germiest Desk

School teachers had about 90 times more germs on their phones than publicists. But the teachers were finally vindicated when their doorknobs were tested for bacteria. They had the cleanest doorknobs of all, with almost 70 times fewer germs than the most contaminated doorknobs. Who had the germiest doorknob? I'll give you a hint: it's the occupation that had the germiest desk and pen.

Who had the germiest desk?

a) accountants

b) bankers

c) doctors

d) radio DJs

The Germiest Desk

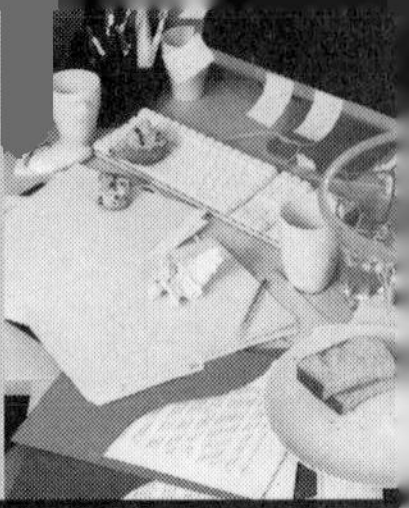

Who had the germiest desk?

a) accountants

b) bankers

c) doctors

d) radio DJs

CORRECT ANSWER:

a) accountants

The accountants' desks, pens, and doorknobs were the germiest. It may not come as a surprise to you that the teachers' desks were almost as bad. Radio DJs had the third most contaminated desks. You'd think that doctors, who see sick people, would have the germiest desks, but the accountants' desks harboured about seven times as many germs.

The Dirtiest and Cleanest Mouse in the House

When The Germinator, microbiologist Dr. Chuck Gerba, and his colleagues tested the computer mouse of working professionals, no one was shocked by the discovery that the average teacher's was the most contaminated. What about the other end of the scale?

Who had the least contaminated computer mouse?

a) bankers

b) doctors

c) publicists

d) television producers

The Dirtiest and Cleanest Mouse in the House

A

Who had the least contaminated computer mouse?

a) bankers

b) doctors

c) publicists

d) television producers

CORRECT ANSWER:

d) television producers

Television producers had the cleanest mouse, with almost 30 times fewer germs than an average school teacher's mouse. But that's their only claim to least-germy fame. Their desks had about two and half times as many bacteria as a doctor's desk. Perhaps it's because television producers are more likely to eat at their desks than doctors. Despite having the fourth most contaminated desks, TV producers had the fourth least germy job overall.

Button Pushers

Once upon a time, office workers were called pencil pushers. These days, it would be more accurate to call them, and most of the rest of us, button pushers. You might push dozens of buttons before you even leave home. It's no secret that frequently touched buttons have lots of bacteria on them. But which one is the germiest and which is the least germy?

Q

Which button is likely to be the least contaminated?

a) bank machine

b) elevator

c) palm pilot or cell phone

d) photocopier

Button Pushers

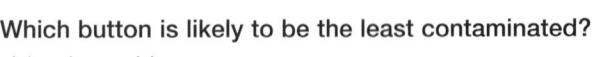

Which button is likely to be the least contaminated?

a) bank machine

b) elevator

c) palm pilot or cell phone

d) photocopier

CORRECT ANSWER:

b) elevator

Elevator buttons don't even make the top 10 of the most contaminated things in your office. The most contaminated elevator button was always the one for the floor everyone used to exit the building. Chinese researchers found an average of 1,200 germs on each key of bank machines. The Germinator, Dr. Chuck Gerba, found 4,000 bacteria per square centimetre (0.15 sq in) on cell phones, which is about 500 times more than on an average toilet seat. For the top 10 most contaminated things in your office, check out *What Does the Moon Smell Like?* (thebraincafe.ca).

Q

It's party time! You've laid out yummy snacks, including crackers and dip. You notice that one of your friends is dipping his cracker, taking a bite, and then dipping again. You normally wouldn't share saliva with this person but if you dip now you might. You know that every time he double dips, he's transferring his spit and bacteria to the dip.

How many double dips transfer 10,000 bacteria to the dip?

a) 1–2

b) 3–6

c) 7–10

d) more than 10

Double Dipping

How many double dips transfer 10,000 bacteria to the dip?

a) 1–2

b) 3–6

c) 7–10

d) more than 10

CORRECT ANSWER:

b) 3–6

Whoever dips after the double dipper gets about 50 to 100 of his mouth bacteria. Luckily, no one has to share anyone else's germs or saliva if they don't want to. Thick dips cling to crackers and chips better, so less spit and germs get transferred into the dip. Crispy snacks that are too small to dip twice are the ultimate solution, unless you have a double dipping friend who takes freakishly tiny bites.

Breaking the Ice

One of your friends has brought their very attractive cousin along to the party. You want to go over and start a conversation. You only have one shot at making a good first impression. What should you say? Good question. Maybe the results of a British speed dating study can give us a clue about the most successful opening lines.

What was the best line from the top-rated female speed dater?

a) Are you accepting applications for your fan club?

b) I have a Ph.D. in computing.

c) If you were a pizza topping, what would you be?

d) My best friend is a helicopter pilot.

Breaking the Ice

What was the best line from the top-rated female speed dater?

a) Are you accepting applications for your fan club?

b) I have a Ph.D. in computing.

c) If you were a pizza topping, what would you be?

d) My best friend is a helicopter pilot.

CORRECT ANSWER:

c) If you were a pizza topping, what would you be?

Does that sounds as corny to you as it does to me? Let's think of it as an example of a line that makes the other person feel light-hearted and comfortable enough to talk about themselves. That's the most important thing for making a connection. Dr. Richard Wiseman (quirkology.com) also revealed in this speed dating study that guys have to be on their toes. Almost half of the women were hyper-speed daters. They made up their mind about a man within 30 seconds of meeting him.

Struttin' Yo Stuff

Some of your friends are bustin' moves. Can guys attract girls by dancing? Sure, but it depends on your style. To find out what women like, a former professional dancer turned psychology professor filmed himself dancing in various styles. Dr. Peter Lovatt blurred his own facial features, to keep the focus on the dance moves and not on himself. He asked 55 women to watch and rate how attractive, masculine, and dominant each set of moves was.

Which moves were rated the most attractive?

a) complex and coordinated (Tony Manero in *Saturday Night Fever*)

b) large and uncoordinated (extroverted drunk guy dancing)

c) small and simple (shy guy)

d) strong and percussive (hip hop)

Struttin' Yo Stuff

Which moves were rated the most attractive?

a) complex and coordinated (Tony Manero in *Saturday Night Fever*)
b) large and uncoordinated (extroverted drunk guy dancing)
c) small and simple (shy guy)
d) strong and percussive (hip hop)

CORRECT ANSWER:

a) complex and coordinated (Tony Manero in *Saturday Night Fever*)

You don't have to be a disco dancer. It's just an example of the confident, well-coordinated, medium-sized moves that the women found the most attractive and masculine. Strong, percussive hip hop style moves were rated dominant. Small, simple moves, like shuffling from foot to foot, got the worst ratings. But there's hope for shy guys. Dr. Lovatt says that simply adding random moves here and there looks more confident and attractive.

The Five Second Rule

Argh! You were laughing so hard, you dropped your snack. You've heard of the Five Second Rule. It says that if you drop food on the floor and pick it up in under five seconds, it's safe to eat. Is it true? To answer the question, Dr. Paul Dawson and his students dropped bologna and bread on different types of flooring they'd contaminated with salmonella, bacteria that cause food poisoning.

How long did it take to contaminate the food?

a) less than 1 second

b) just under 5 seconds

c) 6–10 seconds

d) 60 seconds or more

The Five Second Rule

How long did it take to contaminate the food?

a) less than 1 second

b) just under 5 seconds

c) 6–10 seconds

d) 60 seconds or more

CORRECT ANSWER:

a) less than 1 second

The Five Second Rule was busted. Bacteria invaded the bologna and bread in less than the blink of an eye. Four weeks later, when the researchers retested the contaminated surfaces (wood, ceramic tile, and nylon carpet), they found live bacteria on all of them, sometimes even hundreds. As few as 10 of some bad bacteria can make you sick, so eating fallen food can be risky. It all depends on which invisible germs are lurking on the floor.

Blink-Free Memento

Q

The party is winding down. You want to take a picture of everyone still present, but you're worried that you won't get a good one. Someone's always blinking in the group shot! You'll be happy to know that there's an easy way to make sure that you get at least one good photo, even without expensive technology. Say you have six friends framed for the picture . . .

What's the secret to taking a blink-free photograph?

a) ask everyone to yell "Cheeeese!" and then shoot

b) count down aloud from 5 to 1 and then shoot

c) take 2 or 3 pictures

d) take the picture before anyone expects it

Blink-Free Memento

A

What's the secret to taking a blink-free photograph?

a) ask everyone to yell "Cheeeese!" and then shoot

b) count down aloud from 5 to 1 and then shoot

c) take 2 or 3 pictures

d) take the picture before anyone expects it

CORRECT ANSWER:

c) take 2 or 3 pictures

Dr. Piers Barnes, an Australian physicist, worked out a formula for taking a blink-free photo based on the number of people in the picture. Here's a simplified version: divide the number of people in the group by three if the lighting is good, and by two if it isn't. With six people, taking two or three shots gives you a 99% chance of taking a blink-free photo. The formula works for a crowd of up to 20, but with 50 or more people, all bets are off. Asking your friends to yell "Cheeeese!" is optional.

LOST IN SPACE

Sweden's First Satellite

Astronauts have accidentally launched a smorgasbord of stuff into orbit. A glove dropped by Edward White on the first American spacewalk circled the globe for a month. Losing a $100,000 tool bag embarrassed an astronaut in 2008. But it delighted sky watchers who tracked it for almost nine months, until it burned up in Earth's atmosphere. Also presumed to have met a fiery end was Sweden's first satellite, launched in 1966. You didn't know Sweden had a satellite back then?

What was Sweden's first satellite?

a) bottle of Akvavit

b) Hasselblad camera

c) IKEA catalogue

d) Pippi Longstocking doll

Sweden's First Satellite

What was Sweden's first satellite?

a) bottle of Akvavit

b) Hasselblad camera

c) IKEA catalogue

d) Pippi Longstocking doll

CORRECT ANSWER:

b) Hasselblad camera

Astronaut Michael Collins dropped a Swedish-made Hasselblad camera during a spacewalk and it was jokingly referred to as Sweden's first satellite. Among the many objects lost in space are a spatula, pliers, a toothbrush, various nuts and bolts, and a lipstick-sized container of Gene Roddenberry's ashes. The *Columbia* space shuttle carried a pinch of the cremains of the *Star Trek* series creator to space in 1992. When the container finished orbiting and plummeted through the atmosphere, Roddenberry was cremated for the second time.

It's Raining Junk

Some experts think there could be a hundred million bits and pieces of junk in the debris cloud surrounding Earth. Of those, 19,000 large objects, including satellites, are tracked. They range from the size of a double refrigerator to the size of a softball. Half a million smaller pieces, all the way down to bean-sized, and tens of millions of even smaller bits are not tracked. Big or small, whatever goes up must come down eventually.

How often does tracked space junk fall to Earth?

a) once a day
b) once every few days
c) once a week
d) once a month

It's Raining Junk

How often does tracked space junk fall to Earth?

a) once a day

b) once every few days

c) once a week

d) once a month

CORRECT ANSWER: a) once a day

In the past 40 years, an average of one tracked piece of space junk a day has fallen to Earth. As for junk too small to track, no one knows how much falls. Hypothetically, if one piece smaller than a softball but bigger than a bean fell every day, it would take about 1,400 years for all the pieces that size to fall back down to where they came from.

The Oldest Space Junk

The oldest piece of space junk has been orbiting high above the Earth since 1958. *Vanguard 1*, the fourth satellite ever launched, was basically a small test satellite. It didn't carry a lot of sophisticated technology, but it still made space exploration history.

What was special about *Vanguard 1*? It . . .

a) showed that the Earth is a bit pear-shaped
b) was solar powered
c) was spherical
d) all of the above

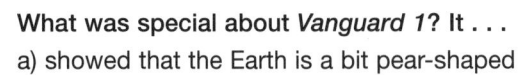

The Oldest Space Junk

What was special about *Vanguard 1*? It . . .

a) showed that the Earth is a bit pear-shaped

b) was solar powered

c) was spherical

d) all of the above

CORRECT ANSWER:

d) all of the above

Vanguard 1 was the first solar-powered satellite. Most of its technology failed within weeks, but the solar-powered transmitter lasted for six years. The satellite's spherical shape was ideal for tracking. By collecting its orbital data, scientists confirmed that Earth is a bit pear-shaped, with the stem end at the North Pole. *Vanguard 1* is basically an aluminum ball about the size of your head that weighs about as much as your brain. But long after your brain is decommissioned, the oldest piece of space junk will still be orbiting the planet.

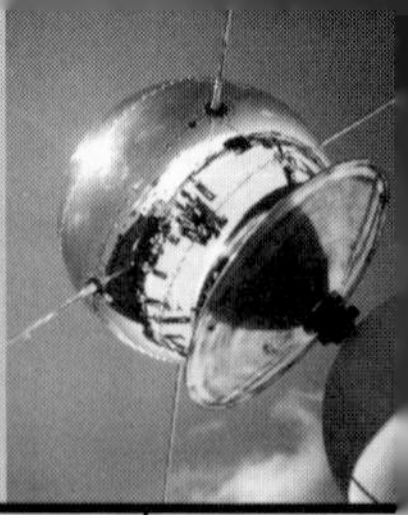

Deadly Dead Satellites

Q

There are close to 900 working satellites orbiting Earth, but they're not alone. Unless it has fallen back to Earth, everything we've launched into space is still up there, including rocket stages, spacecraft parts, broken chunks of electronics, and, of course, non-functioning satellites.

About how many dead satellites orbit Earth?

a) 5,000

b) 2,500

c) 1,250

d) 625

Deadly Dead Satellites

About how many dead satellites orbit Earth?

a) 5,000
b) 2,500
c) 1,250
d) 625

CORRECT ANSWER:
a) 5,000

There are almost five times as many dead satellites as functioning ones. The more junk there is, the higher the risk of accidents and damage to working equipment. In 2007, a satellite had to be boosted out of the way of orbiting junk for the first time, but emergency maneuvers are becoming more common. About 13,000 close calls are tracked per week, and it's a growing concern. If the projections are right, within 50 years there will be more than 50,000 near misses per week.

First Major Collision in Earth Orbit

Q

Dead satellites are supposed to be boosted to the satellite graveyard in the sky, high enough to keep them safely in orbit for hundreds of years longer. But, more often than not, they're just abandoned. Every day, satellites come within a few kilometres of one another. On February 10, 2009, the inevitable happened: the first major collision of two intact satellites in Earth orbit.

At what speed did the satellites collide?

a) about 5,000 km/h (3,106 mph)

b) less than 10,000 km/h (6,214 mph)

c) close to 20,000 km/h (12,427 mph)

d) more than 40,000 km/h (24,855 mph)

First Major Collision in Earth Orbit

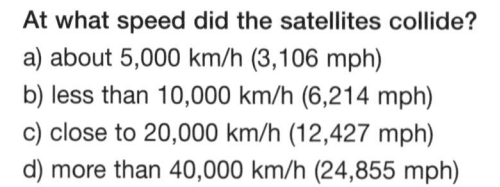

At what speed did the satellites collide?

a) about 5,000 km/h (3,106 mph)
b) less than 10,000 km/h (6,214 mph)
c) close to 20,000 km/h (12,427 mph)
d) more than 40,000 km/h (24,855 mph)

CORRECT ANSWER:

d) more than 40,000 km/h (24,855 mph)

The colossal, high speed impact that shattered *Iridium 33* and *Kosmos 2251* added tons of junk to Earth's debris cloud, but it wasn't the single largest space debris incident. An experiment that exploded an orbiting Chinese satellite in 2007 added thousands of traceable pieces and countless bits and specks of junk to Earth orbit. In less than two decades, the number of tracked pieces of space junk has more than doubled.

TOUR OF THE SOLAR SYSTEM: ATTRACTIONS

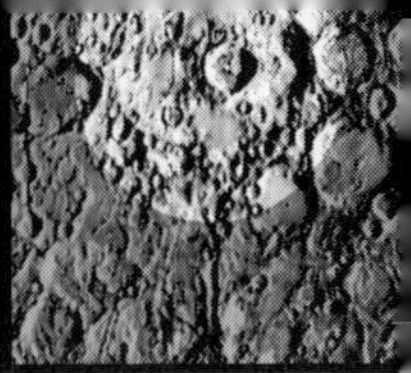

Asteroids Unplugged

Q

You've probably seen a space ship zipping around, narrowly missing rocks in a crowded asteroid field in some sci-fi movie or TV show. The real asteroid belt between Jupiter and Mars is home to billions of space rocks of all shapes and sizes. With so many asteroids, how much space can there possibly be between them? Are sci-fi special effects more science or fiction?

About how far away is an asteroid from its closest neighbour? About the same distance as . . .

a) from New York City to Los Angeles, USA
b) from the Earth to the moon
c) the Earth's circumference
d) the moon's diameter

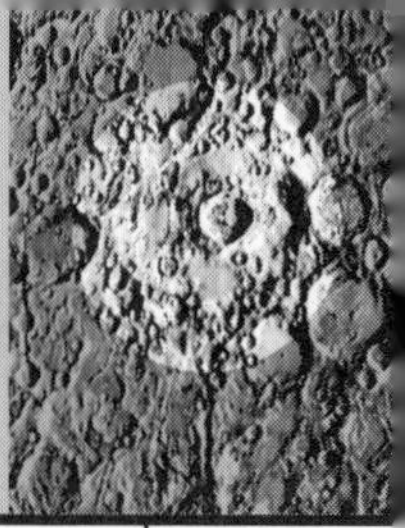

Asteroids Unplugged

About how far away is an asteroid from its closest neighbour? About the same distance as . . .

a) from New York City to Los Angeles, USA
b) from the Earth to the moon
c) the Earth's circumference
d) the moon's diameter

CORRECT ANSWER:
b) from the Earth to the moon

The asteroid belt is so huge that there are likely to be hundreds of thousands of kilometres between an asteroid and its closest neighbour. New asteroids are discovered every day, so the estimates of how many there are keeps changing. But even if there are hundreds of times more space rocks than we think, each asteroid would still have over a million square kilometres to itself. Whipping around between them wouldn't be nearly as hair-raising as it is in the movies.

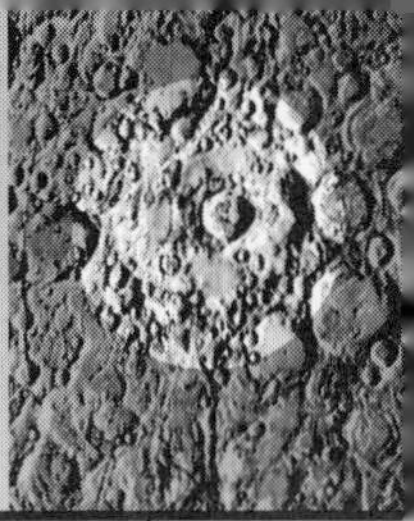

Top 10 Largest Moons

The moon is more than a quarter the size of Earth. It's the solar system's biggest moon relative to planet size. It's so big that some experts think the Earth and moon should be reclassified as a double-planet system. That's an interesting idea, but not likely to happen any time soon. Our moon is relatively big, but how does it compare to the other moons in the solar system? On the top 10 list of the solar system's largest moons . . .

What is our moon's ranking?

a) 1st
b) 3rd
c) 5th
d) 10th

Top 10 Largest Moons

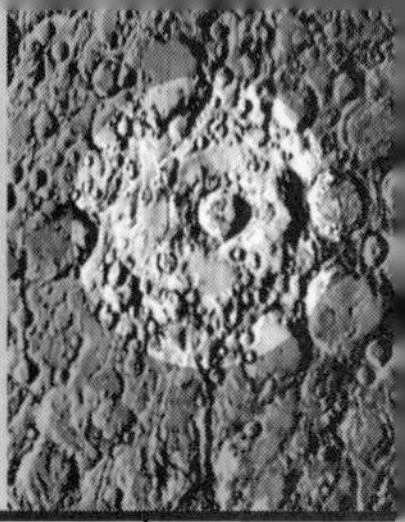

What is our moon's ranking?

a) 1st
b) 3rd
c) 5th
d) 10th

CORRECT ANSWER:
c) 5th

Jupiter's Ganymede is the largest moon, followed by Saturn's Titan, which is the only moon known to have a substantial atmosphere. Ganymede and Titan are both bigger than the smallest planet, Mercury. Jupiter's Callisto has more craters than any other moon and is about the same size as Mercury.

Moon	Mean Diameter *in km/miles*	Planet
1. Ganymede	5,262/3,270	Jupiter
2. Titan	5,150/3,200	Saturn
3. Callisto	4,821/2,995	Jupiter
4. Io	3,643/2,264	Jupiter
5. Moon	3,476/2,160	Earth
6. Europa	3,122/1,940	Jupiter
7. Triton	2,707/1,682	Neptune
8. Titania	1,578/980	Uranus
9. Rhea	1,529/950	Saturn
10. Oberon	1,523/946	Uranus

Einstein Crater

Einstein crater was discovered by an amateur astronomer known to have had musical relations with the genius. After Patrick Moore and Albert Einstein met, their mutual love of music led to a duet, with Einstein on violin and Moore accompanying him on piano. Music aside, Moore's primary passion is astronomy. In 1945, he built a 12.5 inch reflecting telescope, set it up in his garden in England, and started observing. Seven years later, he discovered the crater now known as Einstein.

Where is Einstein crater? On . . .

a) Callisto

b) Europa

c) Moon

d) Titan

Einstein Crater

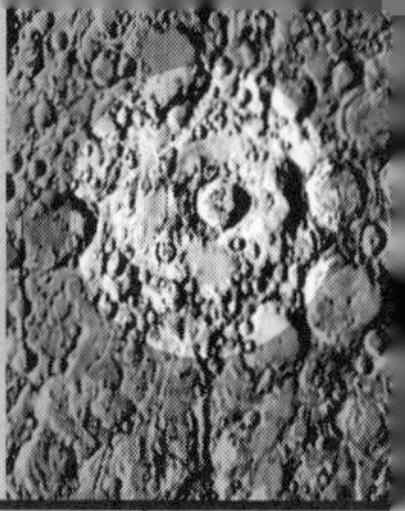

Where is Einstein crater? On . . .

a) Callisto
b) Europa
c) Moon
d) Titan

CORRECT ANSWER:
c) Moon

Patrick Moore started mapping the moon with his homemade telescope. He discovered a large crater at the very edge of the moon, on the western limb. Einstein crater is 198 kilometres (123 mi) across but is rarely seen because of the way the moon rotates on its axis. Since you won't be able to spot it while moon-gazing, have a look at the upper right corner of this page to see the crater. Despite Moore's Einstein connection, he didn't name the crater for him. Originally called Caramuel, it was renamed Einstein by the International Astronomical Union. Do you think they knew about the duet? I'm guessing it's just a really neat coincidence.

The Sun's Magnetic Field

During the solar minimum, when its complexion is clear and free of sunspots, the sun's magnetic field resembles the Earth's. You probably know that the Earth's magnetic field is similar to a bar-shaped dipolar magnet, with open field lines near the ends, or poles, and closed loops near the middle. The sun's dipolar field is like that too. Since the sun is gigantic, it must have a powerful magnetic field, right?

How powerful is the sun's magnetic field? About as strong as . . .

a) a fridge magnet
b) Earth's magnetic field
c) the Large Hadron Collider's field when it's operating
d) the most powerful magnetic resonance imaging scanner

The Sun's Magnetic Field

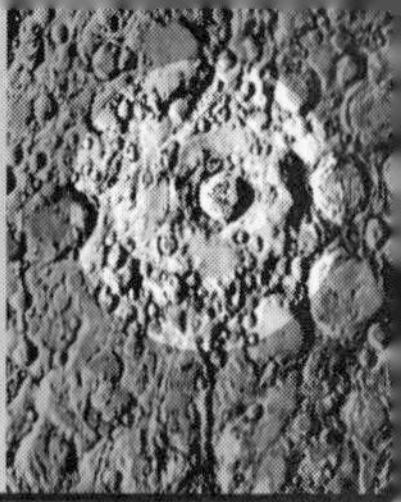

How powerful is the sun's magnetic field? About as strong as . . .

a) a fridge magnet
b) Earth's magnetic field
c) the Large Hadron Collider's field when it's operating
d) the most powerful magnetic resonance imaging scanner

CORRECT ANSWER:

a) a fridge magnet

The sun's dipolar field is about as strong as a fridge magnet, a measly 50 gauss. The similarity to the Earth's magnetic field ends during the solar maximum when huge magnetic loops shoot out of sunspots. They're hundreds of times stronger than the dipole field and overwhelm it. The sun's magnetic field becomes tangled and complicated at the surface, and looks nothing like the Earth's magnetic field.

A Year in the Milky Way

Everything in space orbits something else. The moon orbits Earth; Earth orbits the sun; and the solar system orbits our home galaxy, the Milky Way. I know you know how long the moon's and Earth's orbits are, but do you know how long it takes the solar system to orbit the centre of the Milky Way? It's called a galactic year, and if you know how long it is, please tell the astronomers because they're not exactly sure. Nevertheless, according to their best estimates . . .

What was happening on Earth one galactic year ago?

a) Dinosaurs ruled the planet.

b) Humans had just evolved.

c) Copernicus discovered that Earth orbits the sun, not vice versa.

d) The Hubble Space Telescope was launched.

A Year in the Milky Way

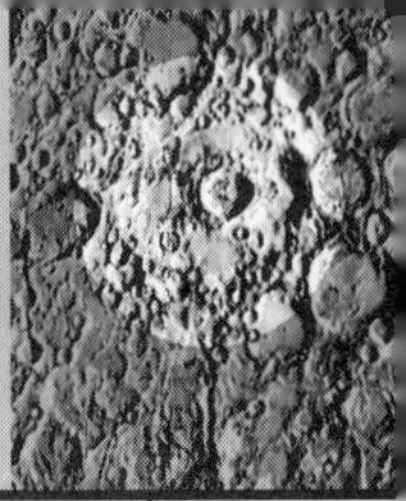

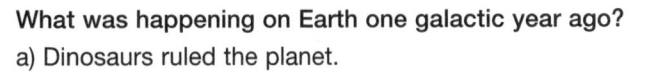

What was happening on Earth one galactic year ago?

a) Dinosaurs ruled the planet.

b) Humans had just evolved.

c) Copernicus discovered that Earth orbits the sun, not vice versa.

d) The Hubble Space Telescope was launched.

CORRECT ANSWER:

a) Dinosaurs ruled the planet.

Experts think it takes the solar system about 225 million years to complete an orbit of the Milky Way's centre. It takes that long due to the vast distances involved, not because we're crawling along. The solar system actually hurtles through space at about 800,000 km/h (500,000 mph). It's hard to imagine anything moving that fast, but it's exactly what you're doing right now. Shouldn't your hair be flying in the breeze? Well, it's the solar system, not a convertible.

Q

In the never-ending quest for a better deodorant, women and men in a Swiss study rode exercise bicycles and sat in saunas while their armpit sweat was collected. Armpit sweat is odourless until bacteria process it, so the researchers mixed some into the fresh sweat and analyzed the resulting chemistry. Then they brought in a panel of professional noses to sniff the stench. In their expert opinion, the female volunteers' sweat smelled like grapefruit or onions.

What did the males' armpit sweat smell like?

a) cheese

b) horseradish

c) pickled cabbage

d) rotten eggs

Women Smell of Onions, Men Smell of . . .

What did the males' armpit sweat smell like?

a) cheese

b) horseradish

c) pickled cabbage

d) rotten eggs

CORRECT ANSWER:

a) cheese

The researchers didn't specify whether they meant Swiss or some other kind of cheese, just that it smelled like pungent cheese. It wasn't nice, but the expert sniffers preferred it to female sweat, which they rated as being the most unpleasant. If your armpits don't smell much like onions or cheese, don't worry. It might just mean that you're not Swiss. Some experts think that factors such as diet, grooming products, clothes, and genetics all contribute to your signature stench.

The Secret Scent of Attraction

Have you ever heard of a white T-shirt experiment? It's used to try to decode the secret messages in personal odours. The experiment involves subjects sleeping in white T-shirts and others smelling and rating the shirts. Do attractive women smell attractive? In one study, women slept in two white T-shirts, one during the fertile stage of their monthly cycle and another at a non-fertile stage. Pictures of the women's faces were rated for attractiveness by a group of male volunteers. Another group of males sniffed the T-shirts, as well as an unworn shirt.

Whose T-shirts smelled most pleasant and attractive to the men?

a) attractive women's
b) average-looking women's
c) skinny women's
d) the unworn ones

The Secret Scent of Attraction

Whose T-shirts smelled most pleasant and attractive to the men?

a) attractive women's
b) average-looking women's
c) skinny women's
d) the unworn ones

CORRECT ANSWER:

a) attractive women's

Without having seen the pictures, the men rated the scent of the T-shirts worn by attractive women and those in the fertile stage of their cycle as being the most pleasant and attractive. Older men were better at sniffing out the fertile beauties than younger guys. Can a few whiffs of a T-shirt really relay that much information? Possibly. Men have preferred the scent of fertile and attractive women in other white T-shirt studies, but no one knows how they do it, or what beauty smells like. It will take many, many more white T-shirt studies to even begin to decode the mysteries of attraction.

The Sneaky Armpit Pad

Q

Young British men slept with pads under their arms to help study the affects of chemical cues, called pheromones, on unsuspecting young women. The females visited a lab on two occasions and sat at a table looking at pictures of men, rating how attractive they were. They didn't know that on one of their visits, a young man's pheromone-laced pad had been taped to the underside of the table. Would the sneaky armpit pad influence their impression of how attractive the men in the pictures were? When the pads were hidden under the table . . .

How did the women rate the men?

a) less attractive

b) more attractive

c) the results varied wildly

d) the same as when there was no pad

The Sneaky Armpit Pad

How did the women rate the men?

a) less attractive

b) more attractive

c) the results varied wildly

d) the same as when there was no pad

CORRECT ANSWER:

b) more attractive

Even though the women couldn't consciously smell it, they rated all the men more attractive when the sneaky armpit pad was under the table. The researchers think that healthy men produce different pheromones than sickly ones, and that women are instinctively attracted to a male's chemical health certificate. With no particular male to associate with the healthy pheromones, the women found all the men in the photos more attractive. Does it work the same way outside the lab, in real life? You'll find lots of opinions and sales pitches, but not a lot of science to answer that question.

Is the Smell of Fear Contagious?

Q

How far would you go for science? Would you jump out of an airplane in a tight neoprene suit with absorbent pads taped to your armpits? Would you do it on your very first skydive? Twenty American volunteers did just that to collect their "fear sweat" for science. The day after their scary jump, they donated more armpit sweat while exercising. Then the researchers scanned the brains of another group of volunteers while they were exposed to the two different kinds of sweat.

What was their response to the fear sweat?

a) activation of the brain's fear centres
b) their brains did not respond
c) thoughts of cheese or onions
d) thoughts of deodorant

Is the Smell of Fear Contagious?

A

What was their response to the fear sweat?

a) activation of the brain's fear centres
b) their brains did not respond
c) thoughts of cheese or onions
d) thoughts of deodorant

CORRECT ANSWER:

a) activation of the brain's fear centers

The volunteers' brains registered fear, but were they feeling it too? The researchers didn't want to influence them, so they didn't ask. But the smell of fear did made the subjects 43% more accurate at judging whether faces in photographs were neutral or threatening. Scientists hope to be able to synthesize the smell of fear some day. With more than 2,000 compounds in body odour, it could take a while. They think that fear sweat could possibly be used to train people with stressful jobs such as soldiers, pilots, or surgeons, or even to keep drivers awake on long trips. Don't we already have caffeine for that?

The Smell of Feet in the Evening

A Dutch scientist became an all-you-can-eat buffet for mosquitoes to learn where different kinds of mosquitoes like to bite. He and his research partner discovered that some kinds had no preference. Others went for the face or shoulders and, most importantly, the mosquitoes that carry malaria went for the ankles and feet. Malaria causes serious illness, and scientists everywhere are trying to come up with ways to lure mosquitoes into traps and away from people. While searching for irresistible mosquito bait the researchers discovered that . . .

Malaria mosquitoes were equally attracted to human feet and . . .

a) bananas

b) Limburger cheese

c) rotting meat

d) wet dogs

The Smell of Feet in the Evening

A

Malaria mosquitoes were equally attracted to human feet and . . .

a) bananas
b) Limburger cheese
c) rotting meat
d) wet dogs

CORRECT ANSWER:
b) Limburger cheese

Once they'd figured out that malaria mosquitoes go for the feet, the Dutch researchers headed to their local cheese shop and loaded up on Limburger, Munster, and other stinky cheeses. They discovered that both stinky feet and cheese that smelled like stinky feet owed their pong to the same bacteria. When Limburger cheese was used as bait, the number of mosquitoes caught in the traps doubled, and even tripled. Malaria mosquitoes couldn't tell the difference between cheese and cheesy feet, a weakness that could be used against them some day.

Q

Armpit odour is bad enough, but when it comes to truly awful stench, does anything beat stinky feet? Ninety percent of the people in one survey thought that men's feet were the worst. Have you ever noticed that? Let's assume that in most cases, it's true.

Why do men have stinkier feet? Because they . . .

a) are less concerned about hygiene

b) don't change their socks often enough

c) eat more junk food

d) sweat more

Why Do Men Have Stinkier Feet?

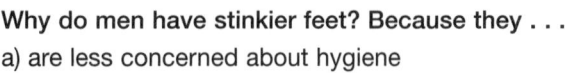

Why do men have stinkier feet? Because they . . .

a) are less concerned about hygiene

b) don't change their socks often enough

c) eat more junk food

d) sweat more

CORRECT ANSWER:

d) sweat more

On the sole of each of your feet, 250,000 tiny sweat glands squirt moisture continuously. Men sweat two to three times more than women. Active feet produce 175 millilitres (6 oz) or more of sweat in a day — each! That's good news for the bacteria that process perspiration and make it stink, but bad news for men with sweaty feet. More sweat plus busy bacteria equals stinkier feet!

The Sexiest Sport

What does the opposite sex find attractive? Do you think you know? To uncover how much people really know about what appeals to the opposite sex, British psychologist Richard Wiseman (Quirkology.com) and fitness guru Sam Murphy teamed up. They surveyed 6,000 people and asked them which sport they thought the opposite sex would find most attractive, and which sport they thought was most attractive in the opposite sex.

What sport did women find most attractive in a man?

a) body building
b) climbing
c) extreme sports
d) soccer

The Sexiest Sport

What sport did women find most attractive in a man?

a) body building
b) climbing
c) extreme sports
d) soccer

CORRECT ANSWER:

b) climbing

Fifty-seven percent of the women picked climbing as the sexiest sport. Did the guys know that? Apparently not. Fifty-six percent of them thought women would be attracted to body building, which didn't even make the females' top 10 list. Women thought running was the sexiest female sport, but it's not what the guys chose. Seventy percent of them thought aerobics was the hottest for a woman. Hmmm. What do you think?

Top 10 Sports for Men According to Women	Top 10 Sports for for Women According to Men
1. climbing	1. aerobics
2. extreme sports	2. yoga/pilates
3. soccer	3. going to the gym
4. hiking	4. running
5. going to the gym	5. skiing
6. skiing	6. hiking
7. running	7. climbing
8. rugby	8. cycling
9. martial arts	9. martial arts
10. rowing	10. extreme sports

The Beauty Bonus

Q

Beautiful people are seen as being smarter, funnier, more athletic, and more socially skilled. Not only do they get more dates, they even make more money than average lookers. Economists who studied the relationship between money and beauty found that in the business world, tall handsome men and attractive women of average weight got a beauty bonus. It could be as much as 41% more money than what average-looking employees earned.

Who was paid the biggest beauty bonus?

a) man in North America

b) man in Shanghai

c) woman in Australia

d) woman in the United Kingdom

The Beauty Bonus

Who was paid the biggest beauty bonus?

a) man in North America
b) man in Shanghai
c) woman in Australia
d) woman in the United Kingdom

CORRECT ANSWER:
c) woman in Australia

Australia had the biggest beauty bonus for both women and men. The penalty for plainness ranged from 1% to 6% in most places, except in North America, which had the harshest penalties. Homely women and men earned, respectively, 14% and 11% less than average-looking employees. What's up with that?

THE BEAUTY BONUS

WOMEN	MEN
Australia 41%	Australia 38%
United Kingdom 37%	Shanghai 32%
Shanghai 34%	North America 30%
North America 32%	United Kingdom 28%

Accounting for Skinny Women

There's a saying, "You can't be too rich or too thin." Busted! You *can* be too thin. Underweight women earn less money than normal weight women. Scrawny women may be financially challenged in business, but some make up for it by marrying money. On average, their husbands made more money than the husbands of other women.

How much more money did the husbands of skinny women earn?

a) 15%

b) 25%

c) 35%

d) 45%

Accounting for Skinny Women

How much more money did the husbands of skinny women earn?

a) 15%
b) 25%
c) 35%
d) 45%

CORRECT ANSWER:
d) 45%

When economists crunched the numbers, they showed that, for some unknown reason, the underweight women in their studies tended to marry men with fat wallets. Did skinny men marry women with big bank accounts? Unknown. But we do know that men weren't penalized for their weight, although short men tended to get the short end of the pay cheque. So far, the trend to reward or punish people based on their appearance has consistently shown up wherever the relationship between money and beauty has been studied.

The Annual Iron Will Challenge

The iron will challenge starts at midnight on December 31st each year. That's when many people resolve to change their bad habits. To reveal the secrets of their success, psychologist Richard Wiseman and his team tracked 3,000 resolution makers. More than half were confident they'd succeed at the start of the study but a year later, only 12% had achieved their goal. How did they do it? That depended on whether the resolution maker was male or female. When trying to lose weight . . .

What two strategies were most successful for men?

a) encouragement from others
b) focusing on the rewards
c) setting firm goals (e.g., losing 1 kg [2 lb] per week)
d) telling friends and family about their resolution

What two strategies were most successful for men?

a) encouragement from others

b) focusing on the rewards

c) setting firm goals (e.g., losing 1 kg [2 lb] per week)

d) telling friends and family about their resolution

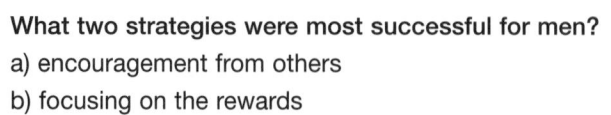

CORRECT ANSWER:

b) focusing on the rewards and

c) setting firm goals (e.g., losing 1 kg [2 lb] per week)

Men who set S.M.A.R.T. goals — Specific, Measureable, Achievable, Realistic, and Time-based — were much more likely to succeed. It also helped if men focused on the potential rewards from being slimmer and more attractive. Women who talked about their resolution to friends, family, and colleagues and received encouragement were almost 10% more likely to succeed. Avoiding previous resolutions and planning a creative approach were some of the other keys to success. But even with all of the above, the hard part remained the same: following through.

Manly Men Eating Popcorn

Q

If a man eats all his popcorn on a movie date, does he seem stronger and more studly? In a quirky Brian Wansink experiment (mindlesseating.org), college students were asked to read one of two stories about a movie date. In version one, the guy in the story, Brad, only ate a couple of handfuls of popcorn. In version two, he ate almost all of it. The male students rated Brad who ate all his popcorn as being stronger. With scientific vigour and a hint of morbid curiosity, the food psychologist asked them how much stronger.

How much more weight did they think Brad could bench press?

a) 1.9 kg (4 lb)
b) 3.6 kg (8 lb)
c) 5.4 kg (12 lb)
d) 9.5 kg (21 lb)

Manly Men Eating Popcorn

How much more weight did they think Brad could bench press?

a) 1.9 kg (4 lb)
b) 3.6 kg (8 lb)
c) 5.4 kg (12 lb)
d) 9.5 kg (21 lb)

CORRECT ANSWER:
d) 9.5 kg (21 lb)

The guys rated popcorn glutton Brad as being considerably stronger and more aggressive. The female subjects weren't impressed with the popcorn eating and rated him about the same as Brad who ate small amounts. The guys thought that eating ravenously was the manly thing to do and bound to impress the girls. Ironically, it only impressed other guys. The males in the experiment were young, but Brian Wansink thinks that manly eating isn't restricted to college students. It's just one of those guy things.

Thinking of Nothing

Try this: whatever you're thinking, say it out loud. Verbalize every thought or word that comes into your head. Weird, isn't it? That's because verbalizing your thoughts alters the thought process itself. Instead of getting people to think about thinking, French scientists tried the opposite approach. They asked their subjects to lie calmly with their eyes closed and think of nothing while their brains were scanned. Have you ever tried thinking of nothing?

What happened when the subjects thought of nothing?

a) After a few minutes they had fewer thoughts.
b) Their minds were flooded with thoughts.
c) Their thought frequency didn't change.
d) They fell asleep.

Thinking of Nothing

A

What happened when the subjects thought of nothing?

a) After a few minutes they had fewer thoughts.

b) Their minds were flooded with thoughts.

c) Their thought frequency didn't change.

d) They fell asleep.

CORRECT ANSWER:

b) Their minds were flooded with thoughts.

Their brains scans showed lots of activity in areas associated with memory, inner speech, emotion, and imagery. The subjects confirmed what their brain scans revealed: 80% reported mental imagery and 75% admitted to talking to themselves in their heads. Their brains were more active when they were trying to think of nothing than during nine demanding mental tasks they'd completed. Apparently, thinking of nothing is harder than solving math problems in your head.

GroupThink

We're bombarded by opinions every day. They come at us from every electronic and communication device, as well as from our friends, families, colleagues, and even complete strangers. How we process opinions and form our own fascinates scientists and master persuaders alike. One study found that when three people in a group express the same opinion, we tend to assume that it's what the entire group thinks. But what happens when one person states their case? If one person repeats his/her opinion three times . . .

How effective is it compared to three people voicing the same opinion?

a) 30% as effective
b) 60% as effective
c) 90% as effective
d) 100% as effective

GroupThink

How effective is it compared to three people voicing the same opinion?

a) 30% as effective
b) 60% as effective
c) 90% as effective
d) 100% as effective

CORRECT ANSWER:
c) 90% as effective

If one person repeats something often enough, others come to think that it's the general opinion. Repetition is the fire that fuels ad campaigns. It sells opinions and products by making them seem familiar. The more familiar something is, the more we're inclined to think that it's the accepted norm. If one persistent voice is 90% as effective as three people voicing the same opinion, what's the effect of the hundreds of repetitious pitches we're exposed to every day? Are we really as immune to them as we think we are?

The Lab Version of Free Will

Q

How do you study free will? Some scientists turned to technology to peek inside a brain in the process of making a decision. In one experiment, volunteers were asked to tap a button with their finger whenever they felt the urge. While they exercised the lab version of free will, their brains were scanned. The scientists compared the timing of activity in brain areas involved in movement to the subjects' conscious decision to move.

What came first?

a) brain activity to initiate movement
b) conscious decision to move
c) finger movement
d) text messaging

The Lab Version of Free Will

What came first?

a) brain activity to initiate movement
b) conscious decision to move
c) finger movement
d) text messaging

CORRECT ANSWER:

a) brain activity to initiate movement

The subjects' brains were initiating movement before they'd consciously decided to move. The lag between brain activity and conscious decision was sometimes as long as 10 seconds. Does that imply you have no free will? Is your conscious mind a slave to your subconscious? Well, whose brain is it anyway? Besides, if your subconscious comes up with a decision or idea that you don't like, you are free to consciously change your mind.

Your Creative Mind

Q

How does the creative mind work? After decades of research, it's still mostly a mystery, though there are a few tantalizing clues. In a neurology study, subjects were asked to invent stories while their brains were monitored. During the first stage of the task, their brains were surprisingly quiet, producing mostly alpha waves, which indicate low activity in the problem-solving part of the brain. During the second stage, the subjects' brains were more active, particularly in the areas associated with conscious analysis.

Who were the most creative thinkers? The ones whose brains showed . . .

a) more activity in the first stage
b) more activity in the second stage
c) same level of activity during both stages
d) the biggest difference in the level of activity between the two stages

Your Creative Mind

A

Who were the most creative thinkers? The ones whose brains showed . . .

a) more activity in the first stage
b) more activity in the second stage
c) same level of activity during both stages
d) the biggest difference in the level of activity between the two stages

CORRECT ANSWER:

d) the biggest difference in the level of activity between the two stages

The most creative stories came from those whose brains were quietest in the first stage and busiest in the second stage. The researchers think that the quiet brains at the start were plumbing their subconscious for ideas. Once they had them, their conscious brain got busy analyzing and evaluating the creative potential of the ideas. The scientists speculate that more subconscious information might spill over into consciousness in highly creative people, giving them richer resources for imaginative connections.

Q

Music is a mind-altering experience that changes your brain. That's the stunning conclusion of a stack of studies that compared the brains of musicians and non-musicians. The musicians played their way to a better brain by "woodshedding" — that's musician-speak from long ago when musicians went out to the wood shed to practice. Playing may be pure pleasure, but it's the hard work of practicing that changes your brain's anatomy. If you woodshed enough . . .

How does music alter your brain?

a) enhances language skills

b) improves reading skills

c) increases grey matter

d) all of the above

Playing Your Way to a Better Brain

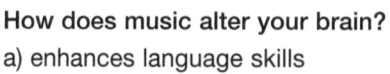

How does music alter your brain?

a) enhances language skills
b) improves reading skills
c) increases grey matter
d) all of the above

CORRECT ANSWER:
d) all of the above

Even if you're just a beginner, woodshedding bulks up your grey matter and literally rewires your brain. The repetition of practicing hooks up more brain cells, forges new pathways, and reinforces existing ones. So, what does this have to do with improved reading and language skills? Music and language share brain areas, and more grey matter and better connections benefit all the functions in those areas.

The Musician's Big Gyrus

For musicians, size matters. The more you exercise your "mental muscle," the more it develops. When scientists compared the brain anatomy of professional musicians and non-musicians, they discovered that musicians had more grey matter in part of a certain gyrus, or convolution, than non-musicians. On average . . .

How much bigger was this gyrus in musicians' brains?

a) 40%
b) 70%
c) 100%
d) 130%

The Musician's Big Gyrus

How much bigger was this gyrus in musicians' brains?

a) 40%

b) 70%

c) 100%

d) 130%

CORRECT ANSWER:

d) 130%

Incredibly, 130% was the average difference. The musician with the most grey matter in his gyrus had 500% more than the non-musician with the least grey matter in the same area. Amateur musicians had less than professionals but more than non-musicians. How much grey matter they had depended on how much time they'd spent practicing music. The good news is that you don't have to be a great player to enhance your brain; all you have to do is practice.

Your Brain on Guitar

Have you ever heard a guitar duet, or duel, that blew your socks off? When the chemistry is just right, it can be an exhilarating experience. To explore the magic of musicians playing together, European scientists took EEG recordings of the brainwaves of eight pairs of guitarists as they played a short jazz-fusion melody in unison over and over again, as many as 60 times.

How synchronized were their brainwaves?

a) very
b) moderately
c) hardly
d) not at all

Q

Your Brain on Guitar

How synchronized were their brainwaves?

a) very

b) moderately

c) hardly

d) not at all

CORRECT ANSWER:

a) very

The brainwaves of musicians who played together stayed together. When they began to play, their brainwaves became virtually identical, oscillating in synchrony from the same areas. Their brains were highly synced in the parts related to music and performance, and, surprisingly, more than half the pairs also synced up in areas linked to higher cognitive functions and subconscious social behaviour. The researchers think these areas may have been activated because of the bond between the players and by their enjoyment of the music.

Plastic Brains

Musicians' brain are plastic — meaning they're flexible, not that they're made of petroleum by-products. Musicians use both sides of their brains more often than the average person. To find out how creative they are overall, an equal number of musicians and non-musicians were recruited for a study. Their intelligence, personality, verbal fluency, and creativity were tested. Then they were asked to come up with new uses for everyday objects.

What did the study reveal about the musicians' brains?

a) They came up with more uses for the objects than non-musicians.
b) They had higher IQs.
c) They scored higher in tests of creativity.
d) all of the above

What did the study reveal about the musicians' brains?

a) They came up with more uses for the objects than non-musicians.
b) They had higher IQs.
c) They scored higher in tests of creativity.
d) all of the above

CORRECT ANSWER:
d) all of the above

It's no coincidence that musicians have been used as a model of brain plasticity in experiments for decades. They're better at thinking outside the box. Musicians came up with about 30% more ideas for the objects than non-musicians, and their ideas were more creative. They also scored higher on IQ and creativity tests. Other studies have also linked higher IQ scores with intensive musical training. The big unanswered question is: does training make musicians smarter or do smarter people pursue music? Or could it be a bit of both? This is a hot topic in brain science, so stay tuned.

Your Brain on Jazz

Have you ever improvised music? It's magical when creativity meets self-expression and goes with the flow. It's almost as if you're the instrument and the music is playing you. What's going on in musicians' brains when they're improvising? The question intrigued scientists and musicians Charles Limb and Allen R. Braun so much, they recruited six jazz musicians and scanned their brains while they improvised.

What were the improvising musicians' brains doing?

a) censoring ideas
b) nothing out of the ordinary
c) planning action
d) telling the musician's personal story

Your Brain on Jazz

What were the improvising musicians' brains doing?

a) censoring ideas

b) nothing out of the ordinary

c) planning action

d) telling the musician's personal story

CORRECT ANSWER:

d) telling the musician's personal story

Brain areas that let self-expression flow were lit up, as were regions involved with all the senses. Brain areas linked to inhibition and self-censoring were turned off. The researchers think that the players shut down impulses that might stifle their ideas and get in the way of the flow state. They noticed that brain patterns during improvisation were like ones seen during REM, the dreaming stage of sleep. Is there a connection between dreaming and improvisation? Are musicians deep in the flow actually in a waking dream? Fascinating questions that scientists and musicians are eager to explore further.

Epilogue

So, how did you do? If you find yourself craving more, check out my first book, *What Does the Moon Smell Like?* You can also visit my website for a sample of that book, more fun facts, and interactive experiments. The Brain Café™ is open 24/7. Hope to see you there.

Eva Everything
www.thebraincafe.ca

Image Credits

Brain Matters: Big Brains
Winning brain, © Blaz Erzetic, iStockPhoto.com

Pet Connections
Turtle, © Mykola Velychko, iStockPhoto.com

Top Secret Projects
Wilbur Franks, © Defence Research and Development Canada, 2010. Reproduced with the permission of the Minister of Public Works and Government Services, 2010.

Got Milk?
Grey pigeon, © Oleg Saenko, iStockPhoto.com

Tour of the Solar System: The Spin
Composite rendering, courtesy NASA

Brain Matters: Better Brains
Human brain, © Henrik Jonsson, iStockPhoto.com

Dirty Secrets: At Home
Eight-inch ceramic tiles, © Eva Everything

Your X-Rated Past
X, Meet Y illustration, © Eva Everything

One of a Kind: You
Newborn sleeping, © Kelly Richardson, iStockPhoto.com

Fastest Man Off Earth
Joe Kittinger, courtesy U.S. Air Force

GermYnation: Poop Factor
Pathological bateria, © Monika Wisniewska, iStockPhoto.com

The Flakiness of Being Human
Police K-9 blood hound, © Luis Santana, iStockPhoto.com

Your Body of Water
Picking nose, © Ilya Andriyanov, iStockPhoto.com

Snack Attack: Ice Cream
Big scoop of ice cream, © Debbi Smirnoff, iStockPhoto.com

The Sounds of Science
Listening globe, © Kwest, Veer.com

Back from the Brink: Peregrine Falcons
Male peregrine falcon from a captive breeding program in Alberta, Canada, © Eva Everything

Getting a Grip on Geckos
Tokay gecko, © Eric Isselée, iStockPhoto.com

Space Rocks
Asteroid Ida, courtesy NASA

Choice Words
Word puzzle illustration, © Eva Everything

Tour of the Solar System: Space Oddities
Jupiter, courtesy NASA

Dogliness
Curree and Lulu, © Eva Everything

Cattitude
Sybil and Pip, © Eva Everything

The Elephant Long Distance Calling Plan
Elephant family at sunset, © Andrea Hill, iStockPhoto.com

Natural Wonders
Spider web detail, © Martin Maun, iStockPhoto.com

GermYnation: Hands On
Flu-masked man, © Jon Schulte, iStockPhoto.com

Dominant Beings
Ants and aphids, © Tomasz Pietryszek, iStockPhoto.com

Faster, Harder, Heavier, Higher
X-1 underneath B-29 Superfortress heavy bomber, courtesy NASA

Psych! My Dinner with Brian
Canned meatballs, © Eva Everything

Brain Matter: More of What Matters
Radiologist pointing at MRI, © Erdar Sipahio Lu, iStockPhoto.com

Rodents for Research
White rat stuck in maze with cheese, © RubberBall Photography, Veer.com

Plastic Ocean
Human footprint, Cozumel, Mexico, © Eva Everything

The Sound of Music
K.D. Lang performs at 1988 Amnesty International Conspiracy of Hope Tour, Toronto, Canada, © Eva Everything

Psych! Seeing and Believing
Mind and pill, © Julien Tromeur, iStockPhoto.com

GermYnation: On the Job
Messy desk, © Fancy Photography, Veer.com

Party Time!
Dancing nerd guy, © Ryan Lane, iStockPhoto.com

Lost in Space
Vanguard 1 satellite, courtesy NASA

Tour of the Solar System: Attractions
Einstein crater, courtesy USGS

The Scent of a Human
Young woman, © drbimages, iStockPhoto.com

Of Women and Men
Lovers, © Eva Everything

Psych! Thinking
Thinking Hurts, © Eva Everything

The Musician's Brain
Playful Data, © Eva Everything